Homo Psychicus as Human and on Becoming a Person

FSC
www.fsc.org
MIX
Papier | Fördert gute Waldnutzung
FSC® C083411

Kolibri 360 GmbH
Ferdinand-Jühlke-Straße 7
99095 Erfurt, Deutschland
produktsicherheit@kolibri360.de

Homo Psychicus as Human and on Becoming a Person

Towards a theory of the human being between the analogue and the digital

Alf Nilsson

PETER LANG
Oxford • Bern • Berlin • Bruxelles • New York • Wien

Bibliographic information published by Die Deutsche Nationalbibliothek. Die Deutsche Nationalbibliothek lists this publication in the Deutsche National-bibliografie; detailed bibliographic data is available on the Internet at http://dnb.d-nb.de.

A catalogue record for this book is available from the British Library.

Library of Congress Control Number: A CIP catalog record for this book has been applied for at the Library of Congress.

Cover image: Created by John Nelander.

Cover design by Brian Melville.

ISBN 978-1-78874-545-1 (print) • ISBN 978-1-78874-546-8 (ePDF)
ISBN 978-1-78874-547-5 (ePub) • ISBN 978-1-78874-548-2 (mobi)

Published by Peter Lang Ltd, International Academic Publishers,
52 St Giles, Oxford, OX1 3LU, United Kingdom
oxford@peterlang.com, www.peterlang.com

This publication has been peer reviewed.

Printed in Germany

To
Ingrid
Jan-Olof
Mats-Ola & Catharina
with Jonathan,
an outstanding source of life during the writing of
my earlier books in Swedish and the present book in English

While life, and indeed all biological and chemical activity, depends upon the ability to combine and transform, the stability of matter and the richness of chemistry is because some things do NOT combine.

Frank Close, *Lucifer's Legacy: The Meaning of Asymmetry*, 2000/2013.

Contents

Figures

Preface

We are living in the Anthropocene epoch. This entails that we as human beings have a profound knowledge of ourselves as biological beings. In addition, we have acquired a comprehensive knowledge of the physical world. At the macro level we have gained knowledge of the origin of universe through the Big Bang, 13.7 billion years ago, and of the growing universe with its galaxies at a distance of several billion light years away. At the micro level, we have gained knowledge of particles at the quantum level, and here our capability for determinacy ends.

It is really astonishing that we as living creatures on earth – with earth as a planet in a solar system on the outskirts of the galaxy the Milky Way – have achieved this knowledge of our universe. In what way are we favoured and deserving of such knowledge as earthly beings? And, as a paramount question: in what way have we utilized this knowledge? One answer to that question might be: to develop digital technologies to such an extent that we are on the verge of jeopardizing the conditions for people's life on earth, a condition caused by our over-exploitation of earth's resources with devastating consequences for the biosphere on earth. Given this grave situation, we might consider that the origin of life is located more than three billion years back in time. Less than half a billion years ago, natural selection and evolution developed simple forms of life into increasingly complex organisms in fertile ecosystems. Just two million years ago the first *Homo* creature saw the light of day. *Homo sapiens* ascended about 170,000 years ago in Africa. *Homo psychicus*, as we know him or her today, started the second and thereby worldwide emigration from Africa 70,000–30,000 years ago. Today it is Man who has power over life on earth. What are we actually doing to our existence on earth?

The Nobel Prize laureate in physiology or medicine in 2000, Eric Kandel, began his career as a psychiatrist grounded in psychoanalysis and psychodynamic thinking. In 2012 he published a comprehensive work, *The Age of Insight*, in which he wrote: "Psychoanalysis [...] was developed by Freud as a dynamic, introspective psychology, *a precursor of modern cognitive psychology*" (italics mine). I don't agree with his description of history. When he maintains a few pages later "As [Ernst] Gombrich was to say prophetically: 'Psychology is biology'", he misses what the essential key issues in psychology are.

My premise is that there is a manner in which all things are organized: nature, objects, our world, affects, operative cognition, mental processes etc. If we formulate a systematic way to describe such a manner, we are formulating a theory. There is no theory that does not involve a manner of being. To borrow a term from the philosopher John Searle, one may sum this up by saying that we are moving into "external and internal realism".

Since psychology, just a century ago, became a scientific discipline, a split has divided psychology: on the one hand in the direction of motivation/affects and, on the other, in the direction of operative cognition. After behaviourism, with cognitive psychology as the heir and at the same time an integral part of mainstream psychology, the only decisive change that has occurred is that cognitive psychology has become a part of "cognitive science". That entails a symbiosis with, among others, cybernetics in the form of "connectionism", and semiotics in the form of "cognitive semiotics".

During my sixteen years as professor emeritus, I've had time to immerse myself in "cognitive science", and I've marvelled at the facile relation between cognitive and affect psychology. In the 1990s it seemed to me that there was an increasing interest in Silvan Tomkins's substantial follow-up of Charles Darwin's suggestion of "the innate and universal emotions". For some reason, Tomkins's thinking about the affects was never widely promoted by cognitive psychology. Instead, cognitive psychology aligned itself with the neurobiological research on affects that had its breakthrough towards the end of last century.

This is my first book on *Homo psychicus* in English. In my approach to the description of *Homo psychicus*, I consider him or her as both an affective/emotional/feeling being and an operative reflex-sensorimotor being in development towards an operative cognitive being. To consider *Homo psychicus* in a developmental frame of reference, involving evolution and ontogenesis, seems to me to be a fruitful way of approaching him.

In the 1970s and 1980s I wrote and published a few theoretical articles and a book in which I was trying to connect Jean Piaget's view of the child's operative development with psychoanalytic developmental thinking. With the passage of time I came to perceive psychoanalysis as theory, and as a dead end for understanding the human being as both an affective and a cognitive being. It was in the 1990s that I discovered Silvan Tomkins's theory of affects.

It is our strong striving for feeling of same-ness, continuity and "me-ness" that has seduced us into regarding the human mind as melted together into one piece, instead of regarding the human mind as emerging *layer on layer* during the evolution (later as a stroke in ontogenesis). Towards the end of his new book and under the heading "Mind/Brain Networks, Layering, and the Brain", Michael Gazzaniga writes: "Layered architectures are one particular modular architecture. Each layer can be thought of as a module."[1] To that I add another quotation from him: "In the current era of human brain imaging experiments, thousands of scientists are committed to finding the places and/or networks that seem to be more active during certain kinds of cognitive states. All realize, however, that this neophrenology may not be capturing the essence of how the brain does its magic to make us who we are and how we feel."[2]

According to the view presented in the book, it is when the feeling of same-ness, continuity and "me-ness" roughly begins to crumble that more or less severe psychopathologies tend to appear. This is also my starting point when describing, in an epigenetic perspective, *Homo psychicus* as a person.

The translation of the present book is the work of Anna Cabak Rédei and the manuscript has been edited by Alan Crozier. I am greatly indebted to both of them.

1 Michael Gazzaniga, 2015, p. 348.
2 Michael Gazzaniga, 2015, p. 336.

Acknowledgements

During my work with the books about *Homo psychicus*, a number of people have assisted in various ways. I am grateful to Cecilia, Christina, Cristina, David, John, Lars, Leif, Mats, Ronny, and Tommy. Special thanks to Frederic Täckström who designed all the books in Swedish.

CHAPTER 1

Background, purposes and some unlocking keys

> Science always aims outwards and upwards – art always inwards and downwards, psychology always outwards and upwards, inwards and downwards.[1]

The present book is the first one published in English about *Homo psychicus*, but it is the seventh in the sequence after six books previously published in Swedish.[2] When I retired in 2001 from my position as professor in clinical psychology, I dedicated myself to the project writing one, and only one, all-inclusive book about *Homo psychicus*. But the subject of the positive affect "interest-excitement" is a capricious endeavour. Now I know from my experience that when "interest-excitement" gets on track, the affect has a tendency to guide human behaviour – in this case myself as a scientist and as an author – towards both wider and deeper inquiries into the issue.

Homo psychicus knows things of the world as well as of her/himself, and he cares about his or her knowledge. No human being would care about what s/he knows unless s/he felt motivated to do so. I distinguish two motivation systems: drive/needs on the one hand, and on the other nine innate affects, the latter being the human "primary motivation system".[3] But without acting and/or executive mechanisms for the management of the internal presentations of the inside as well as of the outside world, nobody would survive in the human sphere. To our aid we have speech and language. Without written language I would not reach you as a reader, or your world.

In my previous books I distinguish between two lines of development inserted into (i) evolutionary development, (ii) the development of

1 Alf Nilsson, 2018.

2 Alf Nilsson, 2005, 2009, 2011, 2013, 2014, 2016.

3 Silvan Tomkins, 1962, 1963, 1991.

Homo (phylogenesis), and (iii) the development of the child (ontogenesis). These developmental lines are those of affect/internal presentation and of operation. Without, at this early stage, going deeper into the details of the systems and subsystems that are linked to these lines of development, I will present some overarching manifest psychological systems which I have examined: the *bottom-up* systems, which, briefly, consist of innate reflexes later manifested in sensorimotor handling, perception (as distinguished from sensation) and Gestalt formation, cognition and language (spoken, written, logic and mathematics).[4] How these systems function in detail, their connection to the two lines of development, and how potential influences and bridges – dynamics – between them might appear, are some themes that will be brought up later on in this book.

My first book about *Homo psychicus* I dedicated to the derivation and description in detail of the systems and subsystems I differentiate on the two developmental lines. Ultimately, I placed these within the frame, which on the one hand starts with the more-or-less *analogue*, on the other hand with the one that starts with the either-or *digital*. It was only with my fifth book that I approached the complex problem of the mutual dynamic influences between the two systems, which are primarily of a bottom-up character, and to some extent of a top-down character.

The main purpose of the book

The aim of this book is to account for the systems and subsystems connected to the two lines of development, and to examine whether there is any justification for describing the interaction between the analogue and the digital systems and subsystems in terms of dynamics. As an introduction to that examination I will present some keys and notions that will reappear throughout the book.

4 Alf Nilsson, 2005.

Some keys

Already in my first book which dealt with "the mechanisms of defence" (see Nilsson, 1983/1994) I pointed out that the English term "representation" is misleading. In that book I rather argued for the German term *Vorstellung*, which in Swedish translates into "föreställning", a term that implies a more correct order than "representation" if one regards what the whole thing is about. This is because we do not know much about the *Vorstellung*, which precedes the *sign/Gestalt*, that is, the *Vorstellung* given a form. Put in Charles Sanders Peirce's terminology, we may categorize the sign in three major classes: icon, index and symbol (see below). On the basis of the link between *Vorstellung* and sign/Gestalt, as outlined here, I have chosen to use the term *internal presentation* here.

The notion of *re-presentation* is used in the book to denominate that something on a lower lever in the evolutionary/phylogenetic and/or ontogenetic system may emerge as something else more advanced on a higher systemic level.

Feedback and *feedback loops* refer to some of the most important characteristics of the central nervous system. The notion of feedback is connected to the notion that what appears in a system influences what has been fed into it. Feedback loops entail a circular flow where that which has fallen out and that which has been fed into a system form a relation of causality.

Strange loop is a flow that achieves a binding of systems on different levels of re-presentation, so that something in the lower system emerges as something else in the higher system. This brings forth, within a feedback loop, that one thing together with another thing reach something new and unique third.

The notion of *metaphoric abstraction* is valid for language when an utterance on a specific linguistic concrete level is expressed as a metaphor on a more abstract level, analogous with what is valid for something in relation to something else when re-presenting.[5]

5 Steven Pinker, 2013.

Semantics is the meaning in (with) language; *syntax* keeps the order so that the right word enters the right place at the right moment.

Expectation (the difference between expectation a priori and a posteriori is addressed later on in the chapter) is a basic element on the operative developmental line – it is a component in the inborn reflexes. However, expectation does not wait for anything specific (object/goal), but waits for anything within certain given frames (a ball of the just right size in the hand to grip, or a cube of the just right size ...), and is without "appraisals".

Sign as a notion has reference to Charles Sanders Peirce. Peirce distinguishes between *icon*, *index* and *symbol*. The icon is based on relations of similarity, the index on contiguity in time and/or space and the symbol rests on rules.[6]

Theme-rheme is the principle that creates intelligibility in an utterance (or text). *Theme* opens the utterance, or the sentence, and includes what is preconditioned to be already *familiar* or *known*. This is what meaning is all about, it is the vantage point for meaning itself. Lev Vygotsky (1896–1934) used the term "psychological subject" for the today established notion of "theme". Vygotsky made a distinction between "psychological subject" and "syntactic subject". *Rheme* is what comes later in the sentence, and which gives *new* additional information to the utterance (or text). Within the Prague School, which was founded in 1926 and had as its most renowned front figure the linguist and literary scholar Roman Jakobson (1896–1982), the contemporary uses of the dichotomy theme-rheme were studied through the relation of "intonation" to "word order". Jacobson's linguistic work follows to a large extent Ferdinand de Saussure's (1896–1982) "linguistic structuralism", with the alteration that linguistic systems and linguistic usages are studied in their respective contexts. The theme-rheme distinction will from now on, throughout the book, be used generally to depict the already known (theme) on the one hand, and on the other hand, that which leads to something new (rheme), from evolutionary, phylogenetic and/or ontogenetic points of view.

6 Charles Sanders Peirce, 1955.

Predicative points forward in an utterance (or in a text) from, for instance, the subject in the sentence to the predicate and/or object. Predicative relates to some feature. Ultimately it is the predicative that points to when, where, how and/or whom it is intended for. To the extent that something in a re-presentation, evolutionarily and/or ontogenetically can be predicted (eventually), I might from time to time refer to a predicative relationship.

Deixis or *deictics* is a term for words or expressions that are placed in a context. One might distinguish between "time deixis", "room deixis", and "person deixis" in the form of pronouns (you, I, he, she, etc.). In the book, deixis/deictics is used occasionally when something points to something else, which leads to a new frame of something and something else.

What is a newborn child?

In the mammal world it only takes a few hours, or at the most a few days, before the newborn offspring can move forward on its four legs. If we consider here the world of our closest primate relative, the chimpanzee, its newborn offspring has a highly developed grasp reflex. Early on the chimpanzee baby has to grasp the mother tightly, in order to take a firm hold of the mother's fur, so as not to lose the grip when the mother swings from one branch to the other, in tree after tree. Under the constant movements of the mother the world of the baby, which is the same as the mother's, takes its shape.[7]

So different things are in the human world. There is no living thing that is so absolutely helpless, so vulnerable and dependent as a newborn human infant, in an absolute sense and 100 per cent during the first year, and then this dependency decreases successively throughout childhood. It is with the arrival of adolescence, adolescence being something unique

7 S. Savage-Rumbaugh and W. Fields, 2011.

to the human progression to adulthood, that an often turbulent liberation (never total) is made possible. During the very first hour there is an essential difference between the newborn human infant and the newborn mammal infant. The child fixates its gaze on the caregivers', who often talk about a gaze contact with elements of "recognition".[8]

By quoting Peter Wolff, I want, early on, to put you on the track of the importance of the difference between these two "worlds", the humans' world and the world of physical things/manufactured objects/artefacts, that this book will be very much about.

> By the end of the first month, there was already a preference for looking at faces over things, [...] By six weeks [...] [l]ooking at the face, particularly when it included eye-eye contact, now captured and held the infant's attention more often than any moving thing displaced across the field.[9]

In the total helplessness the human infant, during the whole of early childhood, differs significantly from the chimpanzee infant. This state of things can be referred to the human infant's premature birth, after a period of only nine months of pregnancy. Richard Leakey comments on the human female pregnancy of nine months as follows: "A simple calculation based on comparisons with other primates reveals that gestation length in *Homo sapiens* [...] should be twenty-one months, not the nine months it actually is."[10]

Within the frames of the evolutionary perspective this particularity is based on *Homo* as a bipedal erect being. In order to achieve a smooth bipedal movement forwards on the ground the *Homo* pelvis had taken on a narrower shape – less suitable for a quick and hasty movement and more so for slow and persistent movement. But this is linked to the shorter period of pregnancy. The narrower birth canal in the pelvis construction did not allow for childbirths of foetuses older than nine months. This also entails that the newborn human, in relation to other primates, still has substantial foetal growth to attain. This requires a motherly world of absolute care and sensitivity, unique to the human world and totally distinguished

8 Sarah Blaffer Hrdy, 2009/2011.
9 Peter Wolff, 1987, p. 205.
10 Richard Leakey, 1994, p. 45.

from that of the other mammals'. Also unique is the nature of the bonding between the individuals in the motherly world and the infant. In the mammals' (and birds') world the bounding may take place at a rapid pace, in a moment, or it may perhaps take up to a few hours or days before an *imprint* is made. In the world of humans, on the other hand, it takes up to two years before any attachment of a lasting character establishes. I have dedicated a large number of pages to describing this complicated process in detail in earlier books.

The equipment of the newborn child

What does the newborn child's equipment look like in order to handle, in the first place, the motherly world and eventually also the outer world filled with physical things/manufactured objects/artefacts which characterize the human world? One of the worlds in which the newborn has to find its way is, as already mentioned, the one where motherly care reigns. For that world, the child is equipped with, if we are to follow in Silvan Tomkins's footsteps, the nine inborn and biologically given *affects*, in this early stage only referred to in their overarching characteristics as being *onefold* or *analogue*. Within the frames of the motherly world, holding prevails, touching skin-to-skin and communication face-to-face, eye-to eye. For the benefit of the continuous development of the child, there is here an abundance of positive affects at the expense of negative affects; to the detriment of the child there is on the other hand an abundance of negative affects at the expense of positive affects.

I will add a section, under a specific heading, that deals with a state of affairs that I have persistently tried to understand during my almost fifteen years of work with my books on *Homo psychicus*. What I refer to here is a proposal of a link between the affects as communicative and what it once, about two million years ago, might have been for the origin of the *Homo* genus, and thereby what it could have meant for the emergence of bipedalism.

The origin of the analogue communication of the affect system

As a first approach to my proposal I use a quotation from Jonathan Kingdon, presented in my first book on *Homo psychicus*. Below follow some parts of it:

> A high proportion of nonadults defines what might have been special about human evolution. It also hints at where the power of natural selection is targeted and is most influential. In thinking that selection favors *smart kids and loving mothers*, I depart from most other students of human evolution, who have concentrated on sexual selection and mate choice in their search for primary selection mechanisms. [...] A social system devoted to nurturing and involving a high proportion of nonadult members is likely to be more tolerant across the board. [...] It is difficult to imagine the group foraging effectively without a great deal of social tolerance and *an appropriate communication system* ... (italics mine)[11]

What Kingdon underlines here is that with the enlarged space for the children, at the expense of the adults', in connection with the origin of the genus *Homo* the former arrived at "the humanizing" of the human being. It is the period of five to six years of pretend play that leads the way to *Homo ludens*. But we still cannot make sense of the evolutionary/phylogenetic origin of, and explanation for the evolutionary "adaptive premises" of the long period of play during childhood for our maturation. My interpretation of Kingdon's underlining of the "stretching" of childhood during phylogenesis, from the origin of the genus *Homo* to the emergence of *Homo sapiens/psychicus*, is that the origins of our humanizing as human beings are to be found in childhood.

As a second driveway to my proposal, which will follow below, I refer to Sarah Blaffer Hrdy's monumental work *Mothers and Others*. Blaffer Hrdy's book is discussed at length in one of my previous books. I've understood Hrdy's work as being throughout a strong argumentation for "cooperative breeding" expressed in terms of *allomaternal/allopaternal* care of the child, from the outset. The prefix "allo" refers to the Greek *allo* meaning "other than" (the mother). I will settle on some short quotations as keys to my proposal for the affect being part of communication (Silvan Tomkins's

11 Jonathan Kingdon, 2003, pp. 294–296.

opinion in his last book was: "The communication of affect is a secondary spin-off function rather than the primary function."[12]) My second key is a quotation from Hrdy:

> [H]uman babies [...] scan the faces of mothers and alloparents not just to predict what they do [...] but to gauge their impressions of what they see, to use their caregivers as "curators of meaning".[13]

Why not end the referencing to Sarah Blaffer Hrdy with a pregnant formulation from her, where she states: "Without alloparents, there never would have been a human species."[14]

Against the backdrop of the above, let us continue with my proposal for analogue communication of innate affects. All decisive things ought to be the inventions of evolution, such as bipedalism and its connection to the origins of the genus *Homo*. Nowhere have I found any discussion about the transition from quadrupedalism to bipedalism and the presumed implications it had for the *Homo* female/mother parturiency. (The *Homo* man ought to have had, with time on his side, a considerably less dramatic adaptation to bipedalism than the *Homo* mother). My proposal considers that the *risk of increased mortality of the* Homo *mother, relative to the newborn child, ought to have been an element in connection with childbirth during the transition era towards the full acquisition of bipedalism.* Those *Homo* mothers who gave birth to children early during pregnancy (foetuses with small heads) ought to have had the best chances to survive. Early births became a strong factor in the natural selection. Why such a critical transition period for the first generation of *Homo* mothers with their instable bipedalism? To answer that question I refer to Richard Leakey's emphasis above of the *Homo* mother's necessary (dramatic?) reorganization towards a pelvis construction adapted to bipedalism, with the implication of increased risks for losses connected to giving birth to foetuses with increasingly bigger heads.

12 Silvan Tomkins, 1991, p. 11.

13 Sarah Blaffer Hrdy, 2009/2011, p. 115.

14 Sarah Blaffer Hrdy, 2009/2011, p. 109.

Sarah Blaffer Hrdy quotes Donald Winnicott: "There is no such thing as a baby. There is a baby and someone".[15] An important predicament that I have been preoccupied with ever since I started my inquiry about the origins and development of *Homo psychicus* is the circumstances for the motherless, thrown-out, and newborn *Homo* child in a *Homo* flock. What s/he had access to in order to communicate his/her emergency to the others in the *Homo* flock was the *innate affects*. As a next step in my proposal I suggested that: *It ought to have been the newborn, motherless* Homo *child who introduced facial expressions for communicating affects (humans have access to 46 muscle groups in the face for such expressions) and in doing so directed the gaze towards someone else in the* Homo *flock asking for (allo)parentship. This way the origin of affects eventually in the form of "interest" and "distress" contributed, on the one hand, to the re-establishing of allomothering and, on the other hand, to the creation of the "first analogue language", a privilege reserved solely for the human and no other animal.*[16] The "directed part" of the affects in the human being is an ontogenetic fundament for, on the one hand, the origin of the affect system as "the primary motivational system" (Tomkins) and, on the other hand it is in the present/the temporally synchronous instant (in German *in dem Augenblick*/in Swedish "*i ögonblicket*") of "specific directed motives" during ontogenesis. These are yet other steps that I will return to later on in the book.

In the communicative perspective the triggering, directed affect takes place in the present/the temporally synchronous and thus is volatile in its character and hard to grasp. As such, there is no possibility to return and erase its expressions afterwards: *the affect as analogue* (valid also for

15 Sarah Blaffer Hrdy, 2009/2011, p. 111.

16 I think it will suffice to quote Sarah Blaffer Hrdy to give the reader an understanding of the term "re-establishing": "In the majority of [mammals], fathers do remarkably little beyond stake out territories, compete with other males, and mate with females. With outlandish auditory and visual displays which often entail specially evolved weaponry, bellowing, barking, or roaring, males engage in fierce contests to rout their competitors." [...] "Yet even though shared care – and, in the case of callitrichids, cooperative breeding – did evolve in many other species than nonhuman primates, none of the Great Apes living under natural conditions in the wild exhibits shared care." 2009/2011, p. 159, p. 230.

the spoken language: the once uttered cannot be unuttered – a residual analogue element in the spoken language?). Only reshaped as a *feeling* may the affect get an extension on a timeline – it obtains the character of being diachronous (the affect reshaped as "emotion" and then eventually as "feeling" and what this means for "corrections" afterwards, is a subject to which I will return later on).

Now to the last element in my proposal for the affects being the incentives of analogue communication at the very outset of the emergence of *Homo* as a species. There is a crucial difference between a biological mother and an allomother in the preconditions in the beginning of the relationship to the newborn child: The biological mother has already during pregnancy initiated a process of attachment to the foetus/the newborn child, something that is not within the allomother's reach. (This is a condition not really considered in contemporary discussions about women's right to have a child via another woman's surrogate motherhood). *It was in an emergency situation of motherlessness and vulnerability that the newborn* Homo *child initiated an appeal, through mobilization of directed affects, to a possible allomother among others in the* Homo *flock.* It ought to have been more natural for a biologic *Homo* mother than for an allomother to form authentic affects in relation to the newborn child. For an allomother, with access to at least embryonic evocative feelings, ought to have a longer journey to make within the self in order to be able to recreate authentic affects in relation to the newborn child. In consideration of this potential situation, the motherless newborn *Homo* child, with his/her (directed) affects ("interest", "distress"), once had to "force" the layers of emotions/embryonic feelings to reach the authentic affects of an allomother with the *Homo* child's directed appeal for care. Is this a phylogenetic relict that is possible to trace in the newborn's widening eyes during his/her first hour of life?

The other world …

The other world is the physical nature with its accumulation of physical things, human manufactured objects, and artefacts. This makes a difference when considering other primates who could easily have contributed to the development of *Homo psychicus*, as we know him or her today as an operative being. To my mind, generally speaking, the production of manufactured objects and artefacts with one technology after another, in an endless stream until our own days, is something that we have not sufficiently considered in order to understand the excessive operative development of *Homo psychicus* (the human being as an "intelligence creature", Jean Piaget). The child, in order to be able to handle this second world initially, has been equipped with the around forty *operative reflexes*. Characteristic for them, on a superior level, is the *twofold* or the *digital*.[17]

Among the child's inborn reflexes we mostly perceive the grasp and suckling reflexes as the most prominent. In my earlier books there are detailed discussions about the reflexes' twofold with the evolutionarily early (at the time of the emergence of the vertebrates) "rhythmic activity" or "biphasic cycle" as the smallest dynamic morphological unit. Early on I had in mind what Graham Brown wrote in 1914 about this in a classical article. Brown provided a compact and exhaustive argumentation against the reflexology-behaviouristic tradition (Ivan Pavlov, John Watson, B. F. Skinner) and its inadequate characterization of spinal reflexes as that which forms the fundamental "morphological unit" of the nervous system for (digital) operative acts of handling. Instead his proposition was "rhythmic activity" which was followed up by the organismic and holistic tradition in psychology.[18] This is how Jean Piaget describes the characteristic feature of "rhythmic activity":

> Whether we are concerned with particular reflex facilitations and inhibitions or, more generally, with a succession of responses in alternating and opposite directions, the rhythm schema always involves, in one way or another, the alternation of two antagonistic processes, the one functioning in the direction A – B and the other in the opposite direction B – A.[19]

17 Alf Nilsson, 2013, 2014.

18 Graham Brown, 1914; Kurt Goldstein, 1939; Heinz Werner, 1948/1957; Jean Piaget, 1947/2001; Paul Schilder, 1964. See also Alf Nilsson, 2005 where I describe the inborn, clearly biphasic, nystagmus reflex as an illustrative example.

19 Jean Piaget, 1947/2001, p. 186.

To this may be added Terrence Deacon's comment, which says that in the biphasic cycle (the modern term) there is an "inherent discontinuity".[20]

In 2011, Teresa Bejarano published *Becoming Human: From pointing gestures to syntax*. The book provides a detailed and rigorous account of both the pre-mental and the early mental functional levels that precede and prompt the development of cognition and language. As the title states, Bejarano primarily wants to trace the emergence of syntax in language.[21]

The author stresses the link between the reflex and what she denominates as "expectation a priori". If we stick to the newborn child's grasp reflex, we may trigger it by placing a ball of just the right size in the hand. It stimulates the sensory sense that conveys signals/sensations to neurons in an area in the cerebral cortex: "F5, a region in the premotor ventral cortex" which is connected through a loop with neurons in the AIP (anterior intraparietal area buried in the intraparietal sulcus IP). These neurons "together are named 'canonical neurons'". Widely applicable, neurons in F5 and in AIP are involved in the grasp reflex, which closes the hand around the ball in a grip without one affectively feeling, perceptually experiencing or cognitively reflecting upon what is happening. Because the canonical neurons with their "vocabulary generate a *basic motor knowledge*", that is, with incentive acts, deal with objects "in the sensorimotor space" – which at the same time excludes the domains for "affects/emotions/feelings, perception, cognitive operation and language".[22]

The reflex act described above resides altogether in expectation a priori. In expectation, completion is included. May one talk about expectation a priori in terms of "motive"? My firm answer is "no!" – with further reference to an extensive discussion in my previous book.[23] Expectation a priori is an element within the frames of the operative developmental line. For expectation the state of affairs is that it might or might not be completed. If expectation is not completed, it will have no part to play in expectation's future possibilities (with the exception of extreme deprivation

20 Terrence Deacon, 2012.

21 Teresa Bejarano, 2011.

22 G. Rizzolatti and C. Sinigaglia, 2006. Quotations from Alf Nilsson, 2009, pp. 99–101.

23 Alf Nilsson, 2013.

of completion). As soon as expectation is involved in the sensation of an object, expectation leaves the object and directs its attention towards the *non-present* object in order to undertake what might come in the next moment. Expectation in the human being ought to have become the necessary "superstructure" over the biphasic cycle, necessary for the handling of the twofold in terms of either the present object or the non-present object, which is the one waiting around the corner in the next moment as an expectation challenge. This way, expectation becomes, in line with Bejarano's description, the foundation for the operative *dualism*, in the long run the *negation*, the *bivalent logic* (Aristotle) and further the *binary difference*.[24] These implications accounted for here are not fully considered in Bejarano.[25]

Expectation as an element on the operative developmental line ought, in its original shape, to have no contact with the affects, which form the fundament for *motivation*. In that respect a *specific difference* ought to prevail between the newborn human child and the newborn chimpanzee child. We have seen earlier that the grasp reflex is a necessary tool from the outset for the infant chimpanzee in order to be able to take a firm hold of the mother's fur in her swings from one branch to the other, in tree after tree. That way, the grasp reflex turns into the essential foundation for the satisfaction of the drives, as well as for the affects – affects in the form valid for the chimpanzee infant. But also, as stated above, the grasp reflex is also fundamental for acquiring knowledge of the outside world (the world of the mother) and eventually for being able to handle the world independently (Figure 1.1). As I have made clear in previous books, the affects became a supplement during the evolution of the mammals (the modest first ones) over 200 million years ago.[26]

Additionally, there is another *specific difference* between the newborn human child and the newborn chimpanzee child. The human child is from the beginning equipped with the Moro reflex, which the chimpanzee

24 Anthony Wilden, 1972/1977. See chapter 3 in this book.

25 Teresa Bejarano, 2011.

26 Alf Nilsson, 2013, 2014.

infant lacks.[27] The Moro reflex is a survival reflex which helps the child to signal to a caregiver that something unexpected occurs: a sharp sound, an abrupt change of the situation, a painful touch, etc. The child first flails out its arms and legs (which would hardly be of assistance to the chimpanzee infant, clinging on to the mother in full swing) and takes a sudden breath. Thereafter the child screams/cries at the same time as it curls up with crossed arms and legs.

Expectation a posteriori

Bejarano makes a distinction between expectation a priori and expectation a posteriori. However, allow me first to state: the emergence of expectation a posteriori is not preceded in any way by of expectation a priori. Instead it emanates in the newborn child's relation to "another caregiver's" *behaviour and kinaesthetic preconditions*, with its origin far down in the "mirror neurons".

The newborn child has access to *mimicking* – which I distinguish from *imitation*, a topic I will return to later – linked to the existence of *mirror neurons* in F5 and AIP. It was Giacomo Rizzolatti and his research group who discovered the mirror neurons during their work of delineating the canonical neurons in macaque monkeys. Rizzolatti takes pain to affirm: "imitation in the true sense of the word is a prerogative of humans [...], but not of the macaques." Furthermore, he states that the mirror neuron system "has an earlier [evolutionary] developmental origin [...] and is primarily involved in the *understanding of motor events*".[28] For instance, the macaque monkey who sees someone else from his own species (or a human being) grasping a nut, something that triggers an *expectation a posteriori* in both the macaque and the human, but only in the latter does it also trigger a motoric behaviour in terms of *mimicking*. This is in line with both Rizzolatti's and Bejarano's opinion. To shed some light on this, here is a quotation from Bejarano, which includes one from Rizzolatti:

27 S. Savage-Rumbaugh and W. Fields, 2011.

28 G. Rizzolatti and C. Sinigaglia, 2006, p. 97. Quotations compiled by the author, from Alf Nilsson, 2009, pp. 100–101.

"parietal-frontal mirror neurons – owing to their motor nature and the fact that they encode the *goal* of motor acts – can be triggered by different visual stimuli that have a common goal (for example grasping)." But with equal, or even greater, clarity they favour my proposal that in mirroring *there would not be latent motor acts, but only a posteriori expectations.* (italics mine)[29]

Figure 1.1. The human mother with the newborn child and the chimpanzee mother with the newborn child, and the respective worlds of the two infants. The picture originally comes from S. Savage-Rumbaugh and W. Fields.[30]

The mirror neuron behaviour of the newborn child is a kinaesthetic prompted motoric act of mimicking without, I repeat, activation of either affective, perceptual or cognitive levels. But the precondition for establishing expectation a posteriori in the human infant is the ability of *mimicking*. In summary, with both expectation a priori and a posteriori – both function within the domain of the operative development – the foundations for extension ought to have been laid: through the starting point in a priori,

29 G. Rizzolatti and C. Sinigaglia, 2010, p. 268; Teresa Bejarano, 2011, p. 33.
30 S. Savage-Rumbaugh and W. Fields, 2011, p. 34.

in an "egocentric" perspective (Piaget) on the one hand, and through the starting point in a posteriori, in an extension towards "altercentrism" and further towards "allocentrism" ("decentrism", in Piaget) on the other hand.[31] Within the frames of a phylogenetic perspective, if we follow Bejarano, "the a posteriori expectation would have originally emerged without any adaptive utility [...] and only afterwards did exaptation take place [with mimicking through the development of *Homo* and his manufacturing of tools]" (brackets mine).[32] One may also describe it by saying that the number of expectations, a priori and a posteriori, as well as the non-completed expectations within humans, has become, during the course of time, endless, or better undefined, insofar as the universe is endless, or undefined. Within the perspective of a human life, the expectations that are triggered by incentives grounded in the affects are the ones that may be *satisfied* or *may not be satisfied*. What is important, within this perspective, that expectations (a priori and a posteriori) may or may not be completed without this predicting future preconditions for the expectations to be triggered.

Rizzolatti and co-workers have pointed out that mirror neurons not only exist in F5 and AIP areas of the human brain. Amongst other things, they prevail in large numbers in the "insula cortex", a formation in the brain evolutionarily contemporary with the emergence of the mammals:

> The information from [...] faces or bodies expressing emotion, is conveyed directly to the insula, where it autonomously and specifically activates a mirror mechanism that immediately codes these descriptions in the corresponding emotive mode. The insula is the centre of this mirror system, not only because it is the cortical region in which the internal states of the body are represented, but because it is the visceromotor integration centre, which, when activated, provokes the transformation of sensory input to visceral reactions.[33]

Louis Cozolina has a detailed account of the role that the insula plays for the human life of affects/emotions/feelings. The insula, sometimes called "the fifth lobe of the brain", is not visible in a fully developed brain. It is covered

31 Jean Piaget, 1977/2001.

32 Teresa Bejarano, 2011, p. 34.

33 G. Rizzolatti and C. Sinigaglia, 2006, p. 189.

by the four other lobes (the frontal, parietal, occipital and temporal lobes), and is situated on the border where the frontal and the temporal lobes meet. From an embryological perspective the insula is the first developed cortex on the surface of the brain – to be superimposed and covered by the other lobes during the embryonic development: "[The insula] appears to play an important role in both the experience of self and our ability to distinguish between ourselves and others". As is suggested in the quotation, the insula is an important transit area for vegetative bodily and sensory experiences. Moreover, it is a structure that plays an important role in the evaluation of facial expressions and gaze, in the process of understanding whether these expressions are indicating positive or negative affects. Thereby the insula is important both for how we learn to separate ourselves from others, and for how we seek companionship with others – for how our "resonance" in interpersonal relations develops: "Utilizing mirror circuitry as a core system and expanding it to include additional systems of social brain, we may be able to begin looking for a fundamental network of resonance and empathy".[34]

The following chapters

Chapter two presents what I am inclined to denominate as a "border zone" between ontology and epistemology, which at the same time establishes a background for the theory of *Homo psychicus*. Thus, a theory about humans on earth and their emergence and development, viewed in a universal and an evolutionary perspective. In Chapter 3, I proceed with an account of the two lines of development, including important steps on the way. The starting point is a schema illustrated in a figure taken from one of my previous books.[35] In Chapter 4, I present a proposal for how the two lines

34 Louis Cozolino, 2006, p. 206, p. 204.
35 Alf Nilsson, 2014.

may bridge and engage with each other, ergo they can be supposed to dynamically interact. They are considered to be the requirement for what leads to the adult *Homo psychicus*. In Chapter 5 I give a summary of what I have written in earlier books about language, this remarkable hybrid of the analogue/semantics and the digital/syntax. Chapter 6 is dedicated to "the person/personality on a clinical ground". After an introduction a description of the psychiatry in Sweden follows. Then follows an account of views of the human becoming a person and a personality and, by extension, descriptions of what characterizes psychopathological states. These states are then illustrated on the backdrop of findings made in studies using PORT, Percept-genetic Object-Relation Test. With reference to an epigenetic perspective, the possible influences of affects, relationships and the dynamics of affects for the development of different psychopathological states are considered. A proposal for a "constraint/interface" as a reference for "warding off" is added. The whole chapter builds on Chapter 5: *Homo psychicus – psychopathologies: affect dynamics, constraints as wardings off, psychotherapy*, in my fifth book in Swedish.[36] In the closing Chapter 7 I present some thoughts, sometime speculative extensions, about *Homo psychicus* as an affective, relational, operative and verbal creature.

36 Alf Nilsson, 2014.

CHAPTER 2

Towards an epistemology forming the background to further understanding of *Homo psychicus*

> As clear as the reference of a sign is the reach out of expectation towards the outside world for completion.[1]

Introduction

At the very end of *On the Origin of Species* Charles Darwin writes:

> In the distant future I see open fields for far more important researches. *Psychology* will be based on a new foundation, that of a necessary acquisition of each mental power and capacity of gradation. Light will be thrown on the origin of man and his history. (italics mine)[2]

In the classic work of Darwin one may distinguish two theoretical premises: one is about the origin with a modification, what we today call evolution. The latter contains the theory of the emergence of simple organisms, which during evolution developed towards more complex beings, ultimately with the human being as the most complex. The other theoretic premise Darwin called natural selection – the primary force that pushed evolution and phylogenesis from the start towards life as we conceive of it today. We are in the midst of an extension and a reformulation of our view of natural selection. Instead of regarding it as a separate independent force, it should be viewed in

1 Alf Nilsson, 2018.
2 Charles Darwin, 1859/2015, p. 576.

the light of interactions, autopoetic developmental and dynamic systems. For this modern view of it the term "evo-devo" (evolution-development) has been coined. In earlier books, as in the present one, I use the term as it stands. But before I take on the task of describing evo-devo, let us start from the beginning.

The Universe, earth and humans in a context of systems theory

In the beginning ...

For my part it started in 1977. That year I published "Adaptive aspects of the individual: A *system approach* to adaptation in relationship to a psychoanalytic anxiety model" (italics mine).[3] At that time artificial intelligence (AI) and cybernetics were the star turns. These were expected to give all answers to the questions regarding the Universe and man on earth. On the level of biopsychology, Ludwig von Bertalanffy had set the first basic tones.[4] My six earlier books have so been a follow-up of the systemic premise.

For a short description of the implications of evo-devo I take as my vantage point three books: one is *Mind in Life: Biology, Phenomenology, and the Sciences of Mind* by Evan Thompson, the other is *Human Development in the Twenty-First Century: Visionary Ideas from System Scientists* edited by Alan Fogel, Barbara King and Stuart Shanker, and the third is *Modular Evolution: How Natural Selection Produces Biological Complexity* by Lucio Vinicius. According to Thompson the new take on the evolution aims:

> to reconceptualize evolution (phylogeny) and natural selection. Evolution is change in the constitution and distribution of developmental systems. Natural selection is not an independent, external agent or force acting on organisms, but rather the outcome of the differential propagation of developmental systems: "It is the net result

3 Alf Nilsson, 1977.
4 Ludwig von Bertanlanffy, 1968.

> of many other interactions in which organism and environment define and select each other" [...]. Natural selection, so conceived, "requires reliable life cycles, not static genetic programmes or organisms [...]."[5]

Further on:

> In the case of animal life, the environment emerges as a sensorimotor world through the actualization of the organism as a sensorimotor being. The organism is a sensorimotor being thanks to its nervous system. The nervous system connects anatomically distant sensory and motor processes, subsuming them in operationally closed sensorimotor networks. Through their coherent, large-scale patterns of activity these networks establish a sensorimotor identity for the animal – a sensorimotor self. In the same stroke, they specify what counts as "other", namely, the animal's sensorimotor world. [...] This is a sensorimotor world – a body-oriented world of perception and action [...].[6]

Now, in my opinion mammals are not only "sensorimotor beings" who lives in what Thompson calls "the sensorimotor world". Earlier I have argued for the view that mammals are equipped with at least "recognitive internal presentations" for recognition of similarities, internal presentations analogous (homologous?) to what Freud denominated as "unbewusste Vorstellungen".[7] There is reason to come back to that.

With a quotation I will give an idea of what Alan Fogel's and his coeditors' view of evo-devo stands for:

> The dynamic systems approach is an emerging interdisciplinary set of principles used by a diverse collection of scientists to help understand the complex world in which we live. The main insight that unites these scientists, despite wide differences in methods and concepts, is a focus on connections and relationships. A relationship between a particular parent and child, for example, is distinguished by the expressions and gestures as well as all the words by which they understand each other. A parent's raised eyebrow might mean "pay attention" or "be careful" to their child. This small and subtle gesture has meaning to both parent and child because they have worked it out together by repeatedly learning how to understand each other, negotiating their mutual needs and goals. The raised eyebrow represents that whole history of the

5 Evan Thompson, 2007, pp. 191–192.

6 Evan Thompson, 2007, p. 59.

7 Alf Nilsson, 1983/1994, 2005, 2009, 2011, 2013, 2014, 2016.

> growth of the relationship. The relationship is a dynamic system because it changes over time (it is dynamic) and because the mutually understood gestures are the result of both people working together to create something that is more than either one of them alone (it is a relationship system). A dynamic system is a relationship that grows over time, has a history, and is more than the simple sum of the parts.[8]

Lucio Vinicius is a modern representative of evo-devo. He does not content himself with starting from the beginning of life. He starts off from the distinction in physics between "determinism" and "probability". With reference to Erwin Schrödinger (1887–1961), who claimed that our reality, or world, consists of particles that only can be defined in relation to each other by probabilistic laws (for example, the electron in relation to the positron), we can distinguish:

> "emergent property" [...] of a complex system of innumerable microscopic parts. He [Bunge] summarised his views in the classic principle "Principle of Order from Disorder", or the origin of macroscopic predictability to microscopic uncertainty.[9]

In Charles Darwin's definition of "natural selection", Vinicius argues, there is a core of thinking concerning the emergence of life and evolution that converges with Schrödinger's thought about "negentropy" (as the opposite pole to the second law of thermodynamics, entropy).

Reductionism has always been used as a weapon against *history*. The difference between physics and organic history lies in which type of *memory* it is about. In the world of physics "memory" is only about "statistical averages". With the organic world in view, Schrödinger suggested a second Principle of Order from Order, where "memory" in a new Order considers "codings" from previous Orders. With the discovery of the DNA structure, Vinicius claims, Schrödinger's second Principle that "selection arises as the result of competition between alternatives" was confirmed.[10] The essence is the view that an element of "progression" is embedded in life from the outset (a thought that Vinicius claims is present already in Darwin).

8 A. Fogel *et al.*, 2008, p. I.

9 Lucio Vinicious, 2010, p. 6.

10 Lucio Vinicious, 2010, p. 10.

If "microevolution" or adaptation is an element due to competition between individuals in a population, then "macroevolution" is a consequence of competition between populations of species (a thought that might also be found in Darwin). A result of this is that more complex species might be formed, where a new more complex Order always contains in its memory "the coding" valid for regulations of earlier Orders. Thus, "historical evolution" is based on the principle for "hierarchical organization" of different "information codings".

About "evolutionary history", Vinicius expresses himself as follows:

> Evolutionary history should be seen as a sequence of historical "tiers of time": the first and lower tier is selection operating at the level of populations; the second is species competition; and the third and most important, represented by episodes of catastrophic mass extinction caused by extraneous factors such as asteroid collisions or radical climate change, results in random sampling of species and the actual pattern of biodiversity that we observe.[11]

The evolutionary version of the Principle of Order from Order is summarized by Vinicius as follows:

> Entities that originally evolved as phenotypic units may gradually adopt a role of information carriers themselves. This has on a few occasions led to the origin of a new type of biological *information coding* for a new form of phenotypic *organization.* (italics mine)[12]

I will conclude my personal description of the "turn of events" (marked with arrows) as a reminder for those who have read my earlier books. For readers, for whom this is the first book, you may regard it as a schematic anticipation of how things might have been in the beginning of the developmental stage to which I will return in detail shortly. Please note that "coding" may be equivalent to "algorithm" in the outline, but not always so.

...**Ribosomes** → **RNA**: on the verge of life → **DNA**: the first of "life codings" with "memory" as "information carrier", with "organization", and with potential for "development" → **Proteins**: from genotype to phenotype,

11 Lucio Vinicious, 2010, p. 17.
12 Lucio Vinicious, 2010, p. 189.

which through information storage of their own and independent behaviour might become genotype → **Acts of handling coding:** on top neurons come with receptivity for patterns of change through learning → **Perceptual coding:** the emergence of stimulus bound *recognitive internal presentations* for recognition of similarities, followed by *evocative internal presentations* as foundation for formation of *images* → **Cognitive coding:** cognitive operations with an element of *reversibility* → **Linguistic coding:** spoken language followed by written language with elements of *recursion* and learning of cultural patterns that eventually can be converted from phenotype to genotype.

I find that the representatives who I have chosen to stand for modern evo-devo thinking have one thing in common. They describe the layers of systems that have emerged in the course of time as "generic emergences" – developments of more complex systems unfolding from less complex systems – in terms of "positive emergences" (which always require positive feedback). To me, there seem to be no sustainable explanations for how anything of lesser complexity can turn into something more complex in the representatives I quote above. Despite different suggestions they have hardly come any further than what Michael Polanyi, in the 1960s, defined as "tacit knowledge". This is valid also for the Gestalt psychologists, when they settled on *description* of the transition from wholes to parts, from ground to figure.[13]

To the account of evo-devo I add the main ideas of the "road map" in Terrence Deacon's recently published book *Incomplete Nature: How Mind Emerged from Matter.* Without doubt, Deacon's "road map" opens for a paradigmatic shift in the way we perceive the relation between the Universe, earth, life and human.

13 Michael Polanyi, 1966/2013.

New favourable directions for evo-devo

The obstacle to an extended modern systemic outlook lies, according to Deacon, in a single-minded positive view of the "self-organizing/regulation". It has constrained further explanations of how "generic emergence" in the course of evolution ultimately gave birth to the human mind. The overall purpose of Deacon's book is to balance with a negative outlook the prevailing positive outlooks of the paradigms forming the generic emergence (see above). With his "constructive" (and not only descriptive) road map, Deacon strives to untie the knot in order to follow the thread that binds together the so far loose and independent fields: nature, life and mind, all dynamic fields. For the author, nature begins with the origin of the Universe.

According to Deacon a modern science worthy of trust can no longer sidestep phenomena such as "values", "functions", "purposes", "meanings", "sentience" etc. In the philosophical and psychological literature these phenomena often end up under the overarching term "intention":

> In philosophical circles, [...] the term *intention* is used differently. [...] It is defined as the property of being *about* something. [...] [T]he common attribute is an involvement with respect to something *extrinsic* and *absent* ... [the two later italics mine] [To avoid repeating all the different meanings, Deacon proposes] the term *ententional* as a generic adjective to describe all phenomena that are intrinsically incomplete in the sense of being in relationship to, contributed by, or organized to achieve something *non-intrinsic.* (these italics mine)[14]

Entention is a neologism derived from "intention". In order to approach such phenomena we need to force the "constraints" and include "absences", or with Deacon's neologism "absentials", which are valid right down to the level of physical material (quantum level). Without constraints the earth would have been a shelter for the second law of thermodynamics, the principle of entropy. What could this have entailed? Constraints of different kinds have made possible, as is known for sure, the development of life and mind. What the results are in terms of absentials, the human can never fully

14 Terrence Deacon, 2012, pp. 26–27.

reconstruct. It is through the way the constraints have been duplicated and multiplied that we may grasp the nature of absentials: they represent merely part of the residue of what was not eliminated. With regard to the human, I claim, we have a unique possibility to reveal the influences of the absentials, from a being born to be human through his conversion to an adult person. My proposal is now that this can be best traced via the development of the human being becoming a person – via epigenesis – where otherwise undistinguishable constraints may have been distinguished in their fixation as "wardings off". To denominate such wardings off, I will come back with terms and notions in Chapter 6.

Before life on earth existed, there was the Universe and the physical nature. With life on all levels that we know of today, "efficient cause" is not enough the way it has been dominating Western natural science's guidelines until today. Deacon widens the notion of causality by going back to Aristotle and makes a distinction between material, efficient, final and formal causes. In my interpretation of Deacon, the Universe constitutes the extreme contextual limit in which, from the perspective of *top-down*, the formal cause – "the absolute algorithm" – is valid. It leads up to three domains included in the road map: from (i) *thermodynamics* (the second law of thermodynamics, entropy), via (ii) "shaped formation and dynamic maintenance", *morphodynamics* (the law of establishing formations), to (iii) "life as being and evolution", *teleodynamics*, in its extreme with *mind* and beings equipped with "*representations*" (cf. my term "internal presentations"). Initially there follows a short summary of the thermodynamics of physical nature (see Chapter 1) and then follows a slightly longer exposition of the morphodynamics of nature.

From chaos via order into disorder

Order may, according to Deacon, be derived from the nineteenth-century physicist and mathematician Ludwig Boltzmann's (1844–1906) view of regularity as grounded "in terms of increasing disorder, thermodynamic entropy". In that way of describing it, only physics and chemistry were valid for earth as nature and order. What we may know about the earthly order before the emergence of life is that transformations took place both continuously (analogue) and discontinuously (digital). With a simple dichotomy "being" may be described as earth with its surface as a border towards the atmosphere. Close to the surface, earth was surrounded by the atmosphere, air filled with steam – and closest to earth was the *biosphere* filled with oxygen, which is a precondition for the emergence of life on earth. We know that steam it may be less or more densified according to the analogue principle of less or more. Densification takes place within given limits. When the densification reaches a certain limit the steam transforms into clouds or possibly rain, according to the principle of material cause – if we consider the clouds as the only "explanation" for the rain. The clouds, in their turn, densified, can thereafter materialize discontinuously and transform into either raindrops or snowflakes. Nowadays we know that snowflakes with one base form, morphology, may take shifting crystal forms in fairytale-like patterns – from thermodynamic to morphodynamic. To transfigure the discontinuity behind the morphology of nature digitally took until 1936. At that time the mathematician Alan Turing (1912–1954) formulated the still valid principle, the algorithm, for the description of the discontinuous, digital or according to the principle of difference, the universal machine, converted into the computer as artefact. The brilliance of Turing's innovation was that he cracked the code for the *momentaneous* and an asynchronous shift – Turing "deprived" the world as "space-time" (Albert Einstein) through the formula for how an electric pulse may be decomposed into a binary code – through "dot/one" or "dash/zero" as signs, and in a *moment* allow them to run on a strip in an endless row. It is important to keep in mind that "dots/ones" and "dashes/zeros" are mathematical *signs*.

Inserted in the evolutionary perspective, language is the first system which is built of, has *emerged* to, symbolic referential signs: The letter combined into words, the words combined syntactically into sentences with a meaning, and thereby language with *both* syntax and semantics. It is worth noticing that the digital sign machine is only controlled by syntax as order (the software giving it the signals/instructions). The digital universal machine would probably still have been virtual if Erwin Schrödinger had not made his important discovery in physics, the distinction between "determinism" and "probability" (see Chapter 1). To repeat: Schrödinger claimed that our reality ultimately consists of particles that may be defined only in relation to each other with the aid of a theory of probability. Nature, to repeat, moves step by step from thermodynamic Order, over the morphodynamic Order to Disorder, entropy at the extreme end; from the emergence of macroscopic predictability to microscopic uncertainty.

This principle may also be converted into the organic reality, life on earth, which, as already mentioned, appeared over three billion years ago. So far nobody has been able to explain with a simple "efficient cause" the transition from matter/energy, via "work" (Deacon's term, see below) towards morphology in nature to the emergence of life. Before I leave the physical world of nature, I shall give you some additional examples of morphodynamic formations in nature. Because of the development of algorithms in mathematics for describing processes in non-balanced (asymmetric) dynamic systems in nature there is today a "language" to denominate the "trajectories" that characterize the processes. In 1963 Edward Lorenz, an American meteorologist, graphically modelled the solutions of equations intended to represent fluid flow behaviour in the atmosphere (Figure 2.1).[15]

Figure 2.2 illustrates three different forms of natural geological polygons transformed by nature through ice formation, ice melting, expansion and retraction.

An early example of "self-organized" formations in nature is what are called "Bénard cells" (Figure 2.3). Bénard cells are formed when a thin liquid stratum is constantly heated from above or constantly cooled from below. This will be my last example of how morphodynamic processes form patterns in physical nature.

15 Edward Lorenz, 1963.

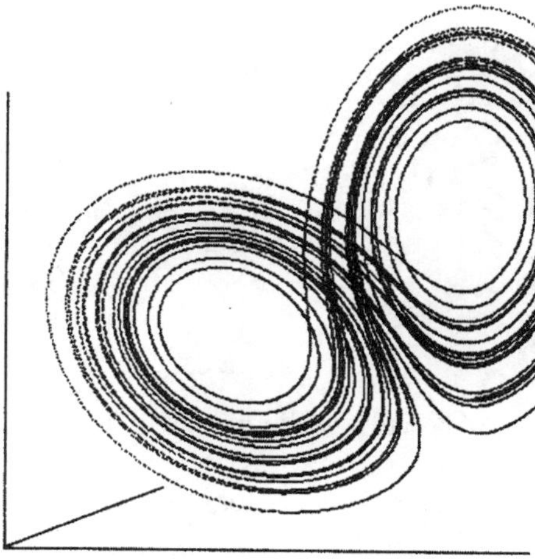

Figure 2.1. Twisted butterfly shape, known as an "attractor", is graphically made (the measurements inserted three-dimensionally) through iterated measurements of simple meteorological processes. The figure is from Terrence Deacon.[16]

Figure 2.2. "Self-organizing" Arctic earth and sand formations (left); regular polygon crack formations caused by ice crystals that have expanded, followed by retractions of earth after the meltwater has dried up (middle); basalt formations in cooling magma layers (right). The figure is from Terrence Deacon.[17]

16 Terrence Deacon, 2012, p. 172.
17 Terrence Deacon, 2012, p. 242.

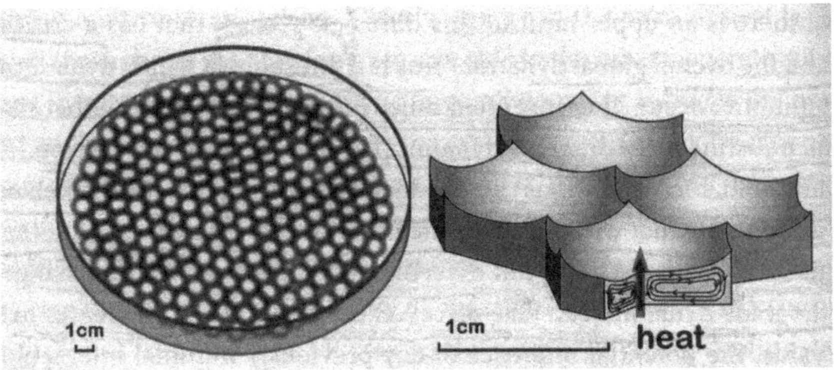

Figure 2.3. Illustration of Bénard cells. The figure is from Terrence Deacon.[18]

Deacon's point in referring to the morphologic formations in Figures 2.1–2.3 is to show that morphodynamic processes exist in physical nature long before they become prevalent elements in teleodynamic formations.

Life, once and for all as teleodynamic

I start with some words about how Deacon integrates the causality forms from Aristotle in his road map. Final or teleological (from the Greek *telos*, "end", "goal", "purpose") cause characterizes all reproductive and replicating life: "means" is something necessary for the survival of the organism, which has reproduction as its "end". Efficient cause, according to Deacon, is treacherous. Via perception as "constantly current force" and "habit" we are exposed to it in our daily life: one thing that causes another. As I have already touched upon, this ought to have contributed to the elevation of efficient cause to an ideal for the Western scientific tradition in its totality, from the natural sciences to the social sciences. And this despite the fact that today it is possible with a scientific outlook to shed some light on complex system formation with help from "discrete mathematics". It takes care of all mathematics, set theory and combinatorics, which deal

18 Henri Benárd, 1900, Terrence Deacon, 2012, p. 250.

with states that may jump from one position to another, which is the case with the digital world.[19]

Is there anything to say about the formal cause? To Deacon, as already stated, the Universe is a dynamic system. The utmost and "absolute" truth about the Universe we may never reach from our earthly horizon. In this matter we are confined to a *bottom-up* conception. In order to reach that truth, as we as humans on earth may come to grips with it, we would need a *top-down* conception. Thus, the acquisition of the ultimate knowledge of the Universe will never be granted to any human being. If we keep to the problem of earth, it is only the human being who has access to the formal cause – everything earthly, from nature to the human, in the *top-down* perspective. This is the premise we are confined to in our study of the emergence of life and evolution/phylogenesis. For that study, according to Deacon, the brain as a whole with possibilities for change may at the best reach "causal probabilities".

As already mentioned, no living creature knows the cause of the emergence of matter/energy. About this we may only form hypotheses. In some rich and detailed chapters, Deacon presents his hypothesis about the transition from inorganic to organic nature. His definition of the missing link is "autogenesis". With that term, Deacon glances at what Francisci Varela denominated "autopoiesis" and Stuart Kauffman "autocatalysis". The lowest common denominator for autogenesis is an excitatory process wired with at least one positive opening and one negative closing feedback loop (which requires a coat/membrane) – at the extreme there is autogenesis characterized by "self-assembly" and "autocatalysis", thus "circular self-regulation". The dynamic processes of autocatalysis and self-assembly are also a source of potential "synergy". On the phenomena of feedback loops, Nicholas Humphrey wrote with insight two decades ago:

19 For a short and popularizing approach to the discrete mathematics, see M. Enquist and K. Eriksson, 2007.

> "Feedback" occurs when the output from a system influences the input to the system; and a "feedback loop" comes into being when, in addition, the input influences the output and a circle of causation is established.[20]

Furthermore:

> Thus what we have, oddly enough, is an overall blending of cause and effect: the As which are the causes of the Bs are also the effects of the Bs. And if, now, we identify this overall long-lasting sequence with the "extended present", we have a situation where the As at the time they go out are already – in the present – under the influence of the Bs they are going to give rise to. Hence the As, from being merely the signals that cause the Bs, have indeed become signals *for and about* the Bs.[21]

In this connection it might be worth bringing up Douglas Hofstadter's inquiries into what he calls "strange loop". It is, seemingly, a paradoxical cross-level feedback loop, in short:

> there are surprising new structures that looping gives rise to that constitute a new level of reality that could *in principle* be deduced from the basic loop and its detailed properties, but that *in practice* have a different kind of "life of their own" and that demand – at least when it comes to extremely finite, simplicity-seeking, pattern-loving creatures like us – a new vocabulary and a new level of description that transcend the basic level out of which they emerge.[22]

According to Humphrey, Hofstadter's "strange loop" may contribute to the understanding of:

> a system when there are several layers of operation, the higher level being built upon the lower, but in which the higher layers are capable of reaching down and changing the structure or rules of the lower levels, [and vice versa].[23]

The microbe (or a precursor to the microbe) appeared as "code" for life on earth, a cell with a membrane, but not necessarily with a cell nucleus.

20 Nicholas Humphrey, 1992/1999, pp. 184–185.
21 Nicholas Humphrey, 1992/1999, pp. 187–188.
22 Douglas Hofstadter, 2007/2008, p. 71.
23 Nicholas Humphrey, 2011, p. 61.

Already for the delimited microbe, there is an interaction with the surrounding nature as ecosystem.[24]

Summary of Deacon's key concepts for a road map of teleodynamics

As it appears above, entention stands behind the human approach towards *extrinsic absential sphere.* Why "entention" instead of "intention"? As I understand Deacon's argument, all living things on earth are restricted to the outer world for the necessary "supplementation" of the *inner* world. Now, my proposal is that "entention" fundamentally is connected to what I define as twofold and human as an operative being (see Chapter 3).

Constraint is close to absence/absential. Constraint impedes certain things and processes from occurring – without Deacon connecting constraint to simple efficacy causality. Even though the author describes constraints which may exist in physical and technical systems, it is primarily how constraints manifest themselves in biopsychological systems that he focuses on. Constraints make it possible for biopsychological systems to convert "energy" into "work", necessary for vital processes, as well as for advanced psychological organization:

> The term *constraint* thus denotes the property of being restricted or being less variable than possible, all other things being equal, and irrespective of why it is restricted. This latter usage has become common in statistical and engineering applications, such as when describing the range of values for some measured feature of a collection of objects, a set of observations, or a sample of informant answers. Used in an intrinsic sense, we might say that the errors in a set of measurements fall within certain constraints or tolerances. [...] In this sense, constraint is a property of a collection or ensemble of some sort, but a negative property. It is a way of referring to what is not exhibited, but could have been, at least under some circumstances.[25]

24 Terrence Deacon, 2012, chapter 10, pp. 288–325; F. Varela *et al.*, 1991; Stuart Kauffman, 1993.

25 Terrence Deacon, 2012, p. 193.

Constraint may achieve dynamic change and due to that produce new constraints, which can be disseminated to adjoining dynamic systems via "the morphological work", because constraints are always "relational". Constraints are connected to formations of *interfaces* (see Chapter 3 below) and thus always consist, in my opinion, of digital elements. One way of describing constraints is as so far undiscovered, more or less analogue formations and digital codes within ourselves, and these have always emerged in relation to the environment. As I understand it, constraints in relation to the environment may be transgressed and discovered by cognitive operation, which has access to both digitalization of interfaces and discoveries of new circumstances in the world/Universe, and of life on earth (evolution) as a consequence, through learning. Peter MacNeilage formulates a proposal for the localization of such a constraint: "For us and our ancestors it is a matter of the body placing *constraints* on the structure of mind, and the mind in turn making demands on the body" (italics mine).[26]

Absentials. Early on in the book Deacon states that: "Longing, desire, passion, appetite, mourning, loss, aspiration – all are based on an *analogous* intrinsic incompleteness, an integral without-ness" (italics mine).[27] They are phenomena that Deacon gathers under the notion of *sentience*, a term that he perceives as synonymous with *feeling*. In my books four and five I direct a certain critique of Deacon on the grounds that he has missed the fact that the human being is equipped with innate affects – present from the beginning of life – for which some of the phenomena he examines can be derived.[28]

To the domain of absentials, as mentioned above, he also addresses: "functions", "purposes", "meanings" and "values". "Our scientific theories still have to officially deny them anything but a sort of heuristic legitimacy."[29]

To absentials, the author also adds figure/background reversal, which, as mentioned above, Michael Polanyi includes in the category of "tacit knowledge".

26 Peter MacNeilage, 2008/2010, p. 134.
27 Terrence Deacon, 2012, pp. 2–3.
28 Alf Nilsson, 2013, 2014.
29 Terrence Deacon, 2012, p. 12.

> A counterintuitive figure/background reversal, focusing on what is absent rather than present, offers a means to repair some of the serious inadequacies in our conceptions of matter, order, life, work, information, representation, and even consciousness and conceptions of value.[30]

According to Deacon, "constraints are the present signature of what is absent" (see above).[31] This may serve as a certain guidance for how "unconscious wardings off" – Freud's generic term in German is "Abwehrungen" – may express itself in PORT, wardings off that otherwise are elements not directly "visible" in psychopathological states. I will return to this subject in Chapter 6.

Deacon rounds off the problem of constraints and absentials with: "There is more here than stuff. There is how this stuff is organized and related to other stuff."[32]

With absentials and constraints there are prerequisites in the human world for: "separation" and "individuation", "me-ness" versus "otherness". This also requires what Jean Piaget defined as "object constancy" and "object permanency", what the psychoanalytic/psychodynamic tradition termed "emotional object constancy" (Margaret Mahler); and on the operative cognitive level "conceptual invariance" (Piaget) (see Chapter 3).

Redundancy. Redundancy is specified by Deacon as the most fundamental notion for room or space. Moreover:

> Redundancy is a defining attribute of dynamic organization, such as the spiraling of galaxies, the symmetries defining geometric forms such as polygons, and the global regularities of physical processes such as are expressed by the second law of thermodynamics. Each is a form of organization in which some process or object is characterized by the repetition of similar attributes.[33]

Habit. The notion of "habit", as illustrated above, Deacon gets from Charles Sanders Peirce. It stands for:

30 Terrence Deacon, 2012, p. 44.
31 Terrence Deacon, 2012, p. 484.
32 Terrence Deacon, 2012, p. 544.
33 Terrence Deacon, 2012, p. 187.

> The general properties by which we assess the similarities and differences among the diverse phenomena of our experience (e.g. pattern and order) are a consequence of the interaction between habits of our brains and habits of the world. So, putting this in constraint terminology, we can dispense with the notion that habit, pattern, and organization presuppose mental representation and comparison.[34]

It is worth noting that it does not have to be about "identity" (at the most "equivalence") between the "habits" of the brain and the ones of nature.

Teleodynamics with morphodynamic processes

Deacon argues for the idea that morphodynamic processes in biopsychological systems could be attributed to a specific form of mutual dependence. This corresponds to what is valid for "attractors" in "cellular automata". Cellular automata are the contrary to the results of the analytic, decomposing starting point in ordinary natural sciences. Simply put, cellular automata consist of a grid with discrete cells. From a separate cell complex patterns of cellular automata develop. This has chiefly been practised in computer simulations. Deacon does not regard computer simulations/ processes as solutions to the problems of teleodynamics: "Computations and cybernetic processes are insentient because they are not teleodynamic in their organization."[35]

If we follow Deacon, cellular automata are also valid for biopsychological systems. The arrangement of attractor cells next to each other is decisive. Deacon does not exclude the possibility that cellular automata may contribute to processes that achieve "internal constraints". Noise in – order out. In my latest books I underline that constraint viewed in this light has similarities with what I denominate as "interface": on the perceptual level in order to delimit "categories"; on the operative cognitive level (level of reflecting abstraction), with language as means, to delimit "classes/ concepts" (see Chapter 3 below). Operative cognition would be a paper

34 Terrence Deacon, 2012, p. 202.
35 Terrence Deacon, 2012, p. 536.

tiger fumbling in a vacuum if it had not been for the fact that the operation has language, and internal presentations, as subject matter to operate with. Moreover, children during epigenesis may be exposed to traumatic afflictions, whereby internal constraints/interfaces may stiffen and become stereotypic and rigid wardings off (Chapter 6).

To repeat: each living organism functions according to the teleodynamic principle. What is common to all life is that it functions according to the principle of "self-interest" in order to survive and reproduce. Deacon uses the term "self" whenever he deals with a living organism – so that on the level of microorganisms the living organism is described as "self-regulating". As is manifested in my earlier books, I am more restrictive in my use of the term "self" – which is used solely with descriptive implications. I will let Deacon conclude: "Teleodynamics can [...] be understood as characterizing the distinguishing dynamics of life [and its morphodynamics that distinguishes it from morphodynamics that crystallize a snowflake in nature]".[36]

Deacon's incomplete nature in retrospect

Deacon considers both what is present and what is absent in his account of what a human being is. It is the absence/absential as "present" which makes it possible for the human being to repeatedly make new discoveries. At the end of his book Deacon writes about his extensive and rigorous study:

> With the realization that specific absent tendencies – dynamical constraints – are critically relevant to the causal fabric of the world, and are the crucial mediators of non-spontaneous change, we are able to stop searching for consciousness "in" the brain or "made of" neural signals.[37]

With the autogenetic creation of "self" as model, Deacon has demystified Descartes' substance dualism by directing attention towards both what is present and what is absent. I understand this to mean that Deacon has tried

36 Terrence Deacon, 2012, p. 275.

37 Terrence Deacon, 2012, p. 534.

to connect the analytic reductionist premise with the one of the synthetic emergent. With regard to the Universe, galaxies, earth in the solar system, the origin of life and evolution, all the way out to the human as a thinking, feeling and morally evaluative and subjectively conscious being, the author has built a platform for "self-reflection" – which ought to be valid for the natural sciences, social sciences as well as for the humanities. In the Western tradition, the systematic thinking took off in the Hellenistic philosophy of Antiquity. By extension, I can see that all scientific activity, in all fields – existing now, as well as those that we have not fully explored, due to internal constraints – includes an element of "philosophical reflection". Then we are heading for a science that, with words from John Searle, adopted by Chris Knight, includes both "brute" and "institutional facts", a science about "everything".[38]

I will allow Deacon to sum up this chapter with some propositions, one which is torn from its context and may appear both condensed and abstract: "Thus, at least for higher-order forms of [human] emergence, the *part/whole distinction and the synchrony/diachrony distinction are intertwined*" (italics mine).[39] As will be made clear in detail in Chapter 4, this proposition may be seen as an entry to how the line of affect/internal presentation is intertwined with the developmental line of operation. Further, a careful formulation and a correction: "I don't claim to have accomplished much more than to have described a hitherto unexplored alternative framing of these enigmatic problems." Moreover: "The subtitle of this book is slightly misleading. Mind didn't exactly emerge from matter, but constraints from matter."[40]

38 John Searle, 1998/1999; Chris Knight, 2009.

39 Terrence Deacon 2012, p. 164.

40 Terrence Deacon, 2012, p. 538.

CHAPTER 3

Two lines of development as the foundation for the origin of *Homo psychicus*

> The human mind did not evolve as an externally guided progression toward either pure reason or emotional fulfillment. It remains as it has always been, an instrument of survival that employs both reason and emotion.[1]

Introduction

As the title of my fourth book in Swedish suggests, translated into English, *Homo Psychicus as Onefold versus Twofold to Manifold*, the foundation for the derivation of *Homo psychicus* is that he follows two lines of development: one rooted in onefold/the analogue; the other in the twofold/the digital. The former I consider to be the developmental line of affect/internal presentation, the latter I describe as the developmental line of operation. Both lines are schematically illustrated in Figure 3.1. The figure clarifies some of the hierarchically organized levels, or systemic delimitations, and additionally some of the important characteristics of both lines appear. Noticeably, the development of the newly born child starts, on the one hand, in the innate affects, and on the other hand, in the innate reflexes. Following Jean Piaget's view, the suggestion is that it is especially the prematurely born human infant who lacks internal presentations – the child is born amental. That there is an early exchange between the fundamental developmental dimensions is marked in the figure with the bidirectional

1 Edward O. Wilson, *The Meaning of Human Existence*, 2014, p. 256.

horizontal arrows. How such a dynamic between the two lines of development might manifest itself will be a subject of its own in Chapter 4, as stated under the heading "Main purpose of the book" in Chapter 1.

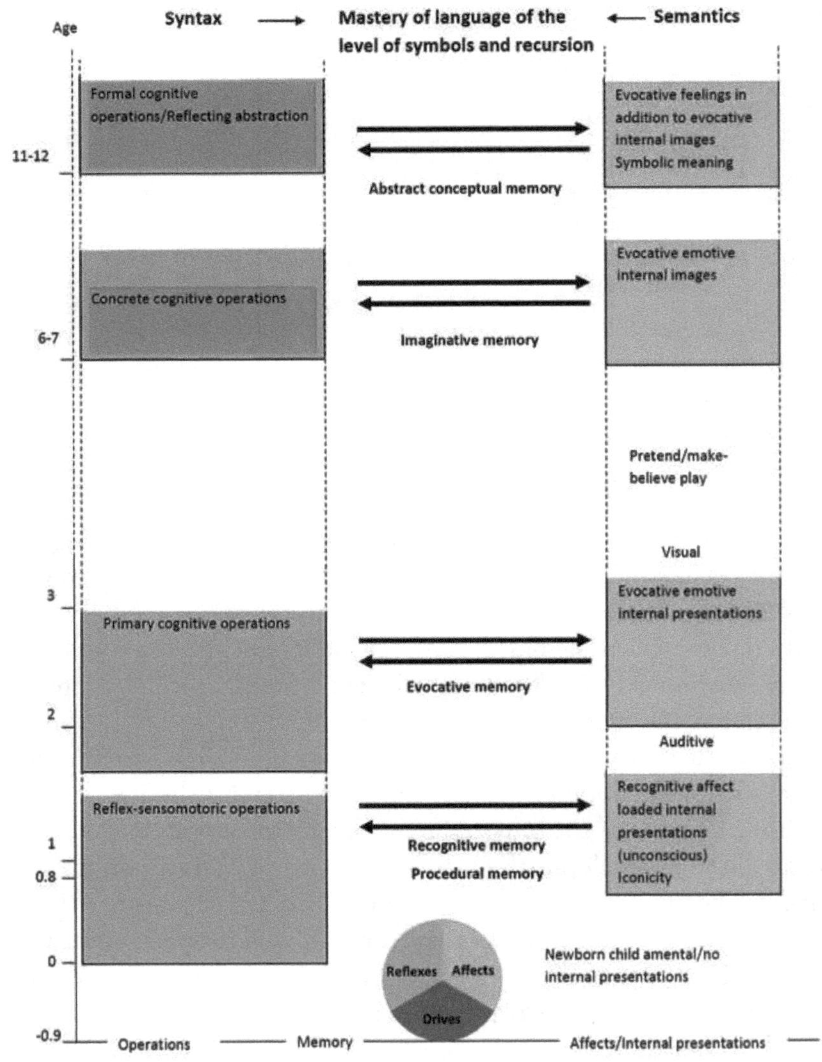

Figure 3.1. Model of the evolution of *Homo psychicus* applied to ontogenesis.

In the scheme the marked arrows connect to different types of memories that are distinguished: procedural, recognitional, evocational, imaginative and abstract conceptual memory. The latter is linked to the process of the development of the child into an adolescent, with its stabilization of the constraints/interfaces of the formal cognitive operations, or the reflecting abstraction and the mastering, through language, of recursively moving back and forth more or less freely between different abstract concepts. It might be appropriate here to give the phenomenon of "information" a more precise definition in the suggested theory about humans and persons.

Information

"Information" can be etymologically derived from the Latin *informare*, "to give shape (to)" or "design". The notion of information has in the course of time been subject to different definitions. With the appearance of the universal digital machine the term has followed the growth of "information science" as a separate branch of the tree of science. Later on in the Chapter I shall enlist Anthony Wilden, Teresa Bejarano and Terrence Deacon as guides in my work of giving precision to the implication of, on the one hand, "information" and, on the other hand, a preverbal "precursor to information", which I will name with a specific term.

First, an epigrammatic definition formulated by the anthropologist Gregory Bateson: "Information is a difference that makes a difference."[2] What I focus on, in this often rendered quotation, with its double meaning of difference, is precisely – "difference". Already in my first book in Swedish on *Homo psychicus* I interlinked "similarity" with "the analogue", "difference" with "the digital". Early on in the book there is a reference to Wilden's classic work *System and Structure*, where Wilden makes a distinction between *analogue communication* and *digital information*. Here follows a more generally formulated quotation from Wilden which touches the quotation from Bateson above: "[A] 'bit' of information is used to select

2 Gregory Bateson, 1972.

one difference from an infinite number of possible differences. This difference then becomes DISTINCT from all the others."[3]

Common to all three of my advisors is the view that information always requires a sender and a receiver. Information arises only when a sign is accepted by the receiver and generates meaning or significance within her or him.

Deacon formulates this in the statement that in *expectation* there is always an "extension towards information". For both Deacon and Bejarano, "information" has a connection with the sphere of "signs", and in this in accordance with Charles Sanders Peirce's definition of signs. Deacon and Bejarano refer to "symbols as signs", basically that information is of linguistic matter. What I've stated earlier to be the absolute criterion for the linguistic symbol is the arbitrary relation, the difference, between the sign/symbol and what the sign/symbol signifies. Basically it may be attributed to the discontinuous, the digital and operative twofold developmental line with its reference in the either-or, the on-off – in the linguistic syntactical recursion.

In the evolution of signs (with the exception of the icon, see below) they are stored in the same development line as expectation. Deacon and Bejarano express the latter idea in terms of expectation reaching out towards the "sphere of signs". Deacon underlines in this context that signs point to "something predictable" about things, physical nature and the Universe. It is a type of predictability that may not be applied to humans. Bejarano expresses this idea by writing that from the beginning expectation occupies itself with what is *not* there – when what is at hand is there (cf. Deacon Chapter 2). From that I draw the conclusion that it is expectation that from the start leads the child to consider the twofold, dualism and difference, eventually to the mastering of notions, and in the long run to the more or less successfully searching for padding and precision of the implications of the notions, in relation to each other.

3 Anthony Wilden, 1972/1977, p. 159.

The developmental line of operation

Let us return to the point where everything starts – the neonate child with the innate reflexes (Bejarano speaks about "innate cognition"). Where the innate reflexes are, there one also finds expectation a priori – of importance for the canonical neurons in F5 and AIP in cortex. As was made clear in Chapter 1, Bejarano separates expectation a priori from expectation a posteriori. To eliminate all doubt, I repeat from what I wrote in Chapter 1 that expectation a posteriori is linked to the infant's immediate *mimicking* of the reactive and kinaesthetic preconditions of the carers – thus, expectation a posteriori may never be preceded by expectation a priori. Described in this way, you as a reader ought to be able to accept the idea that expectation a priori may have an extension towards the domain where "egocentrism" reigns; expectation a posteriori, on the other hand, stretches out towards the sphere of "allocentrism/decentrism". Here I'm reminded about the fact that the child from birth resides in a Universe permeated by language as an auditory phenomenon – but at the same time as visible and kinaesthetic phenomena: mouths and lips are in permanent activity, with a never-ending flow of signs/information, converted into phonetic expressions, pouring out, which the child ought to perceive, more or less – seemingly speechless, note well. Because where the spoken language is, there the infant is not yet. Speechless? Of that we know little. This language buzzes constantly in the environment of the neonate. We still stand puzzled before the question whether the infant is distinctively equipped to encounter and handle language.

Expectation

In the neonate child it is primarily the reflex-sensorimotor equipment that occupies itself with the perpetual buzzing of the language/information that surrounds the child. Today the grasp reflex is inadequate to grip – before the child has caught up with the foetus in order to apprehend. Expectation a posteriori ought then to be especially important in conjunction with this.

But let us look at this from the beginning with expectation a priori. In my previous books in Swedish I have used the example of the "ball, cube, cone". Here I confine myself to the infant, whose first (operative) experience is holding a big enough ball in its hand. This triggers the grasp reflex as an activity, stored in the procedural memory as the "ball". In the example it is arranged in such a way that the child's second experience gets to be a big enough cube in the hand. This prompts the grasp reflex that is tagged "the cube/not the ball" in the procedural memory. Here I draw a parallel with my earlier presentation of the anchorage of the operative developmental line in the twofold in the shape of "rhythmic activity" or the "biphasic cycle" (e.g. Teresa Bejarano, Peter MacNeilage; with certain parallels with what Terrence Deacon calls "bifurcation point") put in more modern phrasing.[4] Twofold in this shaping is characterized by a peripherally prompted activity when either the ball or the cube touches the hand; and a centrally prompted activity which is involved in the imprinting of the "ball" and the "cube" on the procedural memory. The question might arise in the reader: "Is it reasonable to think that the procedural memory operates with negations ("not the ball")?" In my previous books I have argued that this ought hardly to be the case when it comes to the world of chimpanzees and infants. With reference to the instantaneous and precise advancement of the chimpanzee from branch to branch, in tree after tree, "negations" converted into "doubts" could have disastrous consequences for the swinging movements of the chimpanzee in the branches of the trees.

The major difference between the human infant and the newly born chimpanzee is that the latter from the beginning uses the grasp reflex in order to grip and to hold on to the fur of the mother during her whishing from branch to branch … (see Chapter 1). This excludes any preconditions in the chimpanzee child for operating with negations. For the niche of the human child this has led to completely different options. What I have defined, in my thought experiment, as "negations" in the infant's procedural

4 Teresa Bejarano, 2011; Peter MacNeilage, 2008; Terrence Deacon, 2012, p. 249.

memory, Deacon, if I interpret him correctly, defined as "potentials of expectation"; Bejarano as "empty profiles to be filled up in expectation"; what was made clear in Chapter 1 where I denominate "the physical things/ manufactured objects/artefacts" without presence, or "which are *not* there" (as the fundament in the human for operating with negations, this is not made clear in Bejarano).

In the human world, expectation is given completely different preconditions than is the case in other living beings. The question one might ask oneself is whether expectation from the beginning is an element present in the infant's reflex-sensorimotor world (depending on whether the mothering world is caring enough, or not), and as such without influence from the affect system. If the answer to that question is yes, as I presume it is, expectation hardly runs the risk of being circumscribed by "evaluating" restrictions, that is, expectation hardly runs the risk of either starvation or overfeeding. Expectation will continuously always have room for – *redundancy* – for filling, also for unexpected news and information. Where the space of expectation ends, we cannot tell. If it is not endless, it is at least nevertheless undetermined. According to Deacon it might be expressed as follows: "expectation involves a physical difference with a potential towards a physical consequence."[5] I believe that Anthony Wilden is saying something similar when he describes expectation's extension towards information in the following way: "[Expectation in language] is determined [...] by its *syntax* [...], and it is not ruled by causality, but by possibility, *constraints*" (italics mine).[6] I presume that Deacon might agree with this.

Let me continue with my imaginary example. I start where the infant has established, at least, a form of "potential" difference between the "ball and the cube/not the ball" in his or her "reflex-sensorimotor space" – what eventually in ontogenesis implies a passage over the borders for the ball as a *class* towards the cube as a *class.* To study more closely how such a difference of interface might manifest itself, one might proceed in at least two ways. The vantage point for the first way is that the surface of the big enough ball includes an uncountable number of points (actually a subject

5 Terrence Deacon, 2012, p. 394.
6 Anthony Wilden, 1972/1977, p. 35.

for mathematics as language). At this point the procedure of converting the points first into micro quadrangle facets picks up. This is arranged (the child still holds the ball in its hand) in such a way that the size of the facets is increased continuously while the volume is kept approximately constant. The increase in the size of the facets is in this way put into relation to the decrease of the number of facets. My suggestion is that for the child with the physical experiences of both "the ball" and "the cube/not the ball" the psychophysical threshold will eventually be reached when the child leaves (preliminarily denominated) "information" in the reflex-sensorimotor space, the procedural memory, "the ball as the difference in the relation to the cube/not the ball". That "information" is only re-established when the quadrangle facet has reached the status of being the big enough cube/not the ball. What happens to the innumerable number of "facet-in between-modes", that the child has been experiencing until the day the child understands "ball as the difference in relation to the cube/not the ball"? My suggestion is that they are stored as "noise" in the child's procedural memory. As an argument supporting this idea I refer to the fact that the reflex-sensorimotor function is formed in order to experiment with the operation of either-or, either "the ball" or the "the cube/not the ball". Now, as human beings we are shaped to tolerate "noise". That is not the case with the digital machine. As Alan Turing points out: the digital universal machine has to include "redundancy in addition to *reduction* of noise" (italics mine).[7] For each form of "noise" is a threat, working towards the breakdown of the machine.

Now to the second approach. In the imaginary example we have the child with the experience of the "ball". The grasp reflex, as an operation, precedes the experience. From my previous books we know that the inborn reflex is characterized by two congenital preconditions: "flexibility" and "operative efficiency" or "precision". The procedure is now to decrease the size of the big enough ball (the ball still resting in the hand of the child) in the direction of a "small ball". The latter is defined by the fact that it does not trigger any grasp reflex activity – we are now at a level below the

7 Alan Turing quoted in George Dyson, 2012, p. 113.

psychophysical threshold for "the ball" in the "reflex-sensorimotor space". What might still remain is a "small ball" in the sense of being a sensory sensation. In the perspective of the operative grasp reflex I am inclined to denominate all experience of "the ball", before the psychophysical threshold for the "big enough small ball" is reached, as "noise" in the direction of "attunement" (see below). The experiences may, at most, contribute to the actualization of "the potential" (Deacon) or to the filling up of "the empty profile" (Bejarano), that later in life leads to the emergence of "the ball" as a notion, and/or to "roundness". Because, as already mentioned, the processing all the way along the developmental line of operation – from the dormant grasp reflex's first grasping (the big enough ball) to the abstract reflecting of apprehending (the ball/roundness as notions) – is of an either-or type. As an addition: to grasp the big enough ball is to have the ball entirely in one's hand.

Above, I promised to derive a precursor to the possibilities of the formal cognition for operating with "information". If I refer to the thought examples just described, I believe, anyhow, that I might put you on the right track. Let us start by taking a closer look at the linguistic term "information". The first measure I would take is to insert a hyphen between "in" and "formation", in-formation. What appears immediately is that it is about a "formation" that is directed inwards, or has an inward direction. A "formation" is always preceded by a "forming", in the digital machine by a "formatting". Against this backdrop, my suggestion for a precursor of the term "information" is the term "inward formatting" that takes place in the wake of an operational activity on the reflex-sensorimotor level.

I will close this section by reminding you that in all the thought examples involving the operative reflex, expectation is present as a given element: without expectation, no inward formatting of the difference between the "ball" and the "cube/not the ball". However, in the case of the processes I associate with "noise", expectation ought to attenuate (Bejarano) successively and become more disconnected and as an activity end up in the track of what Peirce (and Deacon) describe as "habit". As a rule, I propose that along the developmental line of operations, expectation constitutes the crucial element for the opening towards each new inward formatting as well as for new information. In addition to the problem of information comes the question of the origin of the sign system (icon, index, symbol), ultimately converted into language – the spoken and the written language built on syntax and semantics, and logic and mathematics built on syntax but without semantics in an ordinary sense – as an absolute precondition. Without the sign system, the operative cognition would idle and grope in the dark. To expectation – "expect before" without the expectation anticipating any goal fully – may be added that it never "cares" about filling in or if *no* filling in occurs. It may also be described as expectation being goalless, that is, the expectation has no goal "in sight" (in contrast to intention, to which I will soon return). Thus, expectation is without "morals" and does not "value" inward formatting, nor information. What makes the human "care" about what s/he "knows", that is, what s/he has inwardly formatted, or is informed about, is the inborn affects as the human basic "value and meaning system".

Nine plus twelve months = zero, or one year? (Richard Leakey): Pointing as guidance

The unique human pointing as a *deictic* gesture has been thoroughly studied by Michael Tomasello and co-workers:

> Surprisingly, no one knows where pointing comes from ontogenetically. Specifically, no one knows whether pointing is somehow ritualized by infants from some other behavior, or whether they learn it from others by imitation. [...] [G]iven that some kind of pointing [most common is index finger pointing, but also hand, eye, chin, and lip pointing exist] is very likely universal among human societies, the most plausible hypothesis at the moment is that infants do not acquire their pointing gesture by imitating others; rather it comes naturally to them in some way – perhaps as a *nonsocial orienting action* that becomes socialized in interaction with others (at the beginning only at physical and never at human objects). [...] [F]rom just a few (three) months of age infants have some of the prerequisite capacities for pointing, including the appropriate hand-shape form ... (italics and brackets mine)[8]

8 Michael Tomasello, 2008, p. 112, p. 141.

The latter is a precursor to pointing. It manifests itself in the three-month-old infant keeping the index finger straight while bending the other fingers, sometimes for a long time.

J. Iversen and E. Thelen are bolder in their interpretation when they refer to the embryo in connection with pointing and to the verbal language in connection with the inborn Babkin reflex. Characteristic for the reflex is the fingers and the hand seeking the mouth, apparently goal-oriented (thumb sucking exists already in the foetal stage).[9]

During the last few years I have had the advantage of following Jonathan from birth onwards. In line with Iversen's and Thelen's observations I could see that from the onset the hand and the fingers were seeking the mouth as their counterpart. When Jonathan eventually started to take a grip of things, they always took the short way via the mouth. When he was between five and six months old, I could observe that he first looked with "interest" at his hand and at his fingers as if it/they were outer object(s) – for five to six seconds, sometimes longer – before he put them into his mouth. On this, a commentary by Andy Clark: "First of all, consider the child's own bodily parts. It is quite possible, it seems to me, that these are first experienced [...] as objects in the child's world."[10]

The early connection, via the Babkin reflex, between hand-fingers-mouth – and that intimate linkage, in the continuous development of adult life, between gestures and verbal language – led Michael Corballis to the suggestion that, from an evolutionary perspective, gesture was a forerunner of the verbal language.[11] Iversen and Thelen underline that "communication by gesture precedes, while verbal communication tends to lag behind".[12]

According to Iversen and Thelen, the production of language also has an advantage over the cognitive development. This leads the authors to conclude that language is autonomous in relation to cognition. Steven Pinker expresses the same idea when he writes: "A consequence of the design of grammar [syntax] is that it is a code that is *autonomous* from

9 J. Iversen and E. Thelen, 1999.
10 Andy Clark, 2008, p. 101.
11 Michael Corballis, 2002.
12 J. Iversen and E. Thelen, 1999, p. 33.

cognition."[13] As early as 1979 Elizabeth Bates and co-workers expressed it as: "language is a machine built upon old parts."[14] To that I would add Tomasello's formulation, which at the same time supplements Deacon's "epistemology" that I summed up in Chapter 2:

> Languages have been created within the *constraints* of preexisting human cognition and sociality, and if these are understood well enough, [...] these *constraints* will supply what is needed. (italics mine)[15]

That which Tomasello gives a negative formulation of, I am inclined to turn into a positive suggestion: Language with its syntax and semantics is an extraordinary overlay developed during the course of evolution. It is an overlay existing in *Homo sapiens* that builds upon the *entire equipment* – affects, operative reflex-sensorimotor, recognitive as well as evocative internal presentations, evocative images, operative cognition (Figure 3.1) – that *Homo sapiens* at the time of the emergence of language had at their disposal.[16] In Chapter 5 I will return to this proposal in more detail. But for now I will return to the giant leap, which best describes the child's operative development around the age of one year.

One year old: fivefold rising

In both the cognitive and the phenomenological tradition the incentives for all doing and thinking are *intentions* (from the Latin *intentio* (stretching, straining, tension). As I apprehend it, the evolutionary perspective has been

13 Steven Pinker, 1994/1995, p. 83.
14 E. Bates *et al.*, 1979.
15 Michael Tomasello, 2008, p. 313.
16 In my fourth book in Swedish (p. 112) I emphasize that Adam Kendon is critical of the existence of "a separate module theory" in order to explain the emergence of language. Among other things he says that "modern humans, when they speak together in face-to-face situations, especially in informal settings of everyday interactions, *always* mobilise face and hands and voice together in complex orchestrations". Adam Kendon, 2011, p. 363.

irrelevant to both traditions. Teresa Bejarano, to whom I will soon return, represents an exception. When Peter MacNeilage writes that "intentional control [...] takes place in first words", he indirectly postulates that from the year of one "intentions" may be traced in the child.[17] In Chapter 2 it is clear that Terrence Deacon prefers the term "entention" – founded on something that is also noted by John Searle:

> *Intentionality* is an unfortunate word, and like a lot of unfortunate words in philosophy, we owe it to the German-speaking philosophers. The word suggests that intentionality, in the sense of directedness, must always have some connection with "intending" in the sense in which, for example, I intend to go to the movies tonight. (German has no problem with this because *Intentionalität* does not sound like *Absicht*, the word for intention in the ordinary sense of intending to go to the movies). So we have to keep in mind that in English intending is just one form of intentionality among many.[18]

As is clear from the above, *pointing/pointing gesture* still broods on a secret yet to be discovered, according to Tomasello. Now, it is not only pointing that emerges for the first time at the age of one. The most conspicuous perhaps is the *stand-upright-walk, bipedalism*, which, from an evolutionary perspective, forms, in a commonsense manner among evolutionary researchers, the absolute first step in the human rising towards a linguistic being. In the ontogenetic perspective I have not found anyone who has described a causal relationship between the two skills of pointing and stand-upright-walk. To this there is another supplement to the equipment of the child – the one-year-old who utters the *first word*. As I've presented in previous books, female evolutionary researchers such as Dean Falk and Sarah Blaffer Hrdy agree that the first word exchange took place between a mother and her infant. On the backdrop of a linguistic-technical derivation (of consonants), MacNeilage argues that the first word was "mother", a universal expression but rendered in somewhat different forms in the 7000

17 Peter MacNeilage, 2008/2010, p. 121.
18 John Searle, 1998/1999, pp. 85–86.

languages distinguished today.[19] Moreover – the proposal is Tomasello's, but Bejarano has in all essentials deepened the argumentation that lies behind the proposal – at this age the ability for *imitation* is added, a skill to learn that is necessary in order to acquire language. As this would not suffice – MacNeilage suggests this above, and Bejarano does not hesitate when regarding intention as a further step in (a priori and a posteriori) the intentionality of expectation, which appears at approximately the age of one (Bejarano speaks sometimes of "intention", and sometimes of "desire"). Bejarano's most important foundation for pointing in the one-year-old infant is formulated as follows: "Pointing gestures are always accompanied in children by the communicative use of the eye."[20] Moreover, pointing is associated with the emergence of "the second system" (characterized, among other things, by the ability to "evoke absent objects") – an emergence only reserved for the human being among the primates.

In polemics against Tomasello (who refers to social communication as a motive for the emergence of pointing), Bejarano indicates that for a pointing gesture an access to a motive, *"desire"*, is needed, which is an adaptive asset only if there is an instrumental means connected to the conversion of desire. This postulation may be interpreted to suggest that Bejarano regards the motives as secondary to cognition – and if so, dedicated to what I designate as the "cognitive fallacy". To this I will return later on in this chapter.

To conclude, around the age of one year, important emergences on the developmental line occur: *stand-upright-walk*, utterance of the *first word*, shift from mimicking to *imitation* and conversion of expectation to *intention*; the latter postulates access to efficient as well as final cause (Aristotle). To this comes an observation of Jonathan, thirteen months old: he has just started to toddle about on his own legs – at the same time as he stopped putting new things in his mouth, instead things have only become a matter for grip and gaze.

19 Dean Falk, 2009; Sarah Blaffer Hrdy, 2009/2011; Peter MacNeilage, 2008/2010.

20 Teresa Bejarano, 2011, p. 61.

On the origin of the child's index finger pointing: a proposal

The index finger pointing would likely never have appeared in the child without the innate affect interest's intimate and contingent connection to the gaze and gaze contact – the *pointing gesture* as *deictic*! Note that in this proposal (i) I start with the feature of affect interest, (ii) at least the embryo of visual perception (before visual perception is fully evocative and shaping) as a feature, to end up in (iii) sensorimotor *operative* pointing as feature. Can these three features be put together in terms of a causal link of some kind? As is made clear above, Teresa Bejarano suggests that a pointing gesture in the child is always guided by a communicative element in the gaze; elsewhere: "That practically all pointing gestures are accompanied by alternating looks is, [...] a good clue [in order to trace the origin of pointing]."[21]

A short interlude about affects, eye-to-eye, phylogenesis, "brute", and "institutional" facts

I start with a question: "Does the affectless gaze exist?" Perhaps you answer: "We do speak about the 'empty gaze'." We commonly use the term for patients who suffer from depression. Frankly put, as I see it, behind the "empty gaze" one may always detect the affect of distress. As long as there is the smallest spark of life left, there is always a hint of affect in the gaze.

Before I turn to the objective of approaching something like a causal link with regard to my three connected but with essentially different features – affect, perception, operation – a quotation from Sarah Blaffer Hrdy, in which she gives a time marker for the entry of the modern relational and collective human being into evolution:

> I am convinced that *emotionally modern* humans date back much further. By emotionally modern I mean bipedal apes born with giving impulses and emphatic, intersubjective aptitudes profoundly different from those we see in chimpanzees today – people

21 Teresa Bejarano, 2011, p. 68.

> preadapted to get along with one another even when crowed together on an airplane. Such hominins, I suspect, emerged in Africa hundreds of thousands of years before inventive, symbol-generating, and talkative humans did.[22]

If we take Hrdy literally, the human being was born a human being from his (mammal) affects – and not from his own handmade tools.

What do I actually mean by this? As is apparent in Chapter 1, a contingency system is always a system with a *history* – a system with features that teleodynamically influence each other. We have created the digital universal machine with more and more refined offshoots that allow us to measure and to take further steps towards the macro cosmos as well as towards the micro cosmos – the human is not stupid as an operative cognitive, or intelligent being. Intelligence or operative cognition still have, to our knowledge today, a special kind of human history.

Against operative cognition I put affect, relationship, internal presentation. A primary question: Is the eternal fate of the human to stand fumbling before the affects, elusive in his or her depths as onefold, while already at birth be in contingency with the gaze, and even with the eye-to-eye relationship? We have come far in our knowledge about the world of cells, molecules, transmitter substances and neuropeptides. Next question: Despite this knowledge, will we always stand perplexed before the affect in contingency with the gaze, in contingency with the eye-to-eye relation, at the extreme converted into the visual perception as a "shield"? With the structural and functional perception established one and for all in ontogenesis, the only accessible "language" for perception is *Gestalt and continuity*. Despite the fact that we humans are voyeuristic beings as no other animals on earth, visual perception may never (as diachronous, but never with the achieved range of asynchronous/twofold/operative/digital) *shape* its own origin in the contingence affect → gaze → relationship eye-to-eye. But with this I do not close the doors once and for all. During the course of time Man has shown himself to be the ingenious, sometimes even the brilliant being, always on the lookout for new facts and ideas. But for this we are forever dependent on "language", in one or another form

22 Sarah Blaffer Hrdy, 2009/2011, p. 66.

or shape. This makes the whole a top-down process – but in a way, most probably, via affect/semantics always initiated bottom-up. Inserted in the phylogenetic perspective there is no mystery in this.

The idea came from Claude Lévi-Strauss – but it is Anthony Wilden's merit that the idea has become a bit clearer in its contours. When I took on the idea in my fourth book in Swedish, I started from John Searle's distinction between "brute facts" – cells, molecules, transmitter substances, neuropeptides – and "institutional facts" – established laws about relationships, marriages, ideological views converted into political parties etc. Wilden attaches "brute facts" to "binary differences", "institutional facts" to "binary oppositions", for the latter the task is to formulate the sharpest and most inviolable binary pairs.[23] Will "institutional facts" forever have a special type of human and scientific history than that of "brute facts"? To my knowledge, no firm and walkable bridge has so far been built between the two types of human histories, one of which I call the history of the onefold, the other one I call the history of the twofold. As is made clear in Chapter 2, Terrence Deacon has at least presented a proposal for a possible bridge between the two kinds of facts. This may serve as a background to my proposal for the emergence of finger pointing – so far with an origin shrouded in mystery – which will be foregrounded in the following section.

Finger pointing as an extension of gaze pointing

As far as I know, there are no studies that scrutinize potential connections between gaze and finger pointing. From Michael Tomasello we know that finger pointing has a precursor at the age of three months, when the infant keeps the index finger straight and other fingers bent. At the same age one assumes that the sucking reflex ceases and that the sucking transforms to become more controlled in sensorimotor operative terms. Then the child puts (for instance) a moderately big ball in its mouth and sucks on it. At

23 Claude Lévi-Strauss, 1995; Anthony Wilden, 1972/1977; John Searle, 1995/1996.

this stage the child alternatively sucks on his or her fingers or hand, and the ball, cube, cone etc.

With Jonathan as an example, I describe above how the child between five and six months with "interest", for shorter or longer periods of time, looks at his or her hand and fingers – as if it/they were "physical foreign objects" – before it puts them in its mouth and sucks on them. "Interest"? There might be another way of describing the procedure – as purely operative! If we take Teresa Bejarano as a vantage point the incentive might be expectation – probably not expectation a posteriori that always requires kinaesthetic mimicking of someone other/carer (with the utmost reference in the mirror neuron in F5 and AIP) – thus, with the incentive expectation a priori. Now, my proposal is that without any first embryonic "constancy" in the contingency system affect → gaze → the relation eye-to-eye, expectation a priori would be, in this situation, too volatile and without enough "endurance" for dealing with any observation of the fingers and the hand before they/it are/is put into the mouth. I adhere to the strongest (primary?) incentive being expectation a priori (and not a posteriori). Thus, I propose something that I will follow up shortly, that expectation as an element in the operation has got access to *time* as an operative feature – something that the onefold synchronous, analogue and irreversible affect will never have the good fortune of getting – for the five- to six-month-old infant time developing as a feature in expectation a priori to "put the fingers and hand into the mouth either "in the moment", or "never", time as *moment*. In what follows I suggest treating experience of "instant" "synchronous", "now" and "Augenblick", but *not* "moment" (see below), as synonymous; and "moment" as a time state for operating with.

Only the human child has the good fortune of being able to look at his or her fingers and hands as "other natural objects", and not only as features via "the now/synchronous/onefold" (see Figure 1.1, not even the chimpanzee child comes close to this). This I see as crucial element for the development of time as a feature in the operation – and if you wish, as a "supplement to expectation", in the long run intention – before the "sensorimotor decision" *either in the moment or never*, a time decision (the six-month-old infant has not yet acquired the feature "next moment"). In this connection two questions arise that I think are worth asking: Is this the "key moment" itself, in the ontogenesis of the child, when time becomes a feature in expectation/operation – in order to convert itself later on with the utmost operative expression, from what we know today, through the reflection abstraction, into Albert Einstein's famous formula $E = mc^2$? Can "anything else", primarily in the form of the child's own fingers and hands, emerge in time without the child being anchored in the contingency affect → gaze → relationship eye-to-eye? Note: Affect is never the timekeeper (nor does it measure). If we follow Bejarano it is not the physical thing/ ... which is there (in time), that expectation primarily occupies itself with (but only touches) – but with the physical thing/... as un-expected that is *not* there, and towards which expectation sends out its feelers. When the fingers and the hand of the six-month-old child finally become "other (natural)objects" of the same dignity as physical things/..., the mouth, with its non-nurturing sucking, is both the inner and outer space/time for the child's exploration of it.

When for the child, at the age of around eight months, the physical thing/... reaches the hallmark of "object constancy" (Piaget's term, although I prefer the term "spatial object constancy"), it implies that the child perceives the physical thing/... as "one and the same" (identical or equivalent?) physical thing/... in position A and position B when it is *moved* in the room: the object *here* as the same object *there*. My suggestion is now, because it is not a physical thing/... that is present, but the unexpected one that is *not* there which expectation seeks, the position of A *in relation to* position B in the room, is what occupies expectation (the physical thing/... has started to take on the character of "habit") – and thereby the point in time is also set for *when* time starts to be a concern for expectation (and operation) – at the same time as "the embryo of difference" and "relation between differences" emerge. What does expectation "care" about the nature of the physical thing/... when the given only consists of the fact that it is the unexpected, and when only that not-expected physical thing/... is of concern? This is when/where time enters expectation of/(when) the next unexpected/not-expected physical thing/... will appear in the room. To that we may add that this happens at the same time as *recognitive internal presentation capacity* (according to Piaget's "recognition memory") emerges in the child – the child moves from an amental to a primary mental being, to which I will return below.

Why does time become a "necessary" feature in the child's operative ability? In an environment filled with physical things/manufactured objects/artefacts, entirely different from the one of the chimpanzee child, every physical thing/… takes a "necessary" moment ("In […] 1925 Heisenberg [earmarked] [… among other terms] *momentum* (*p*) […] as an 'operator' on [quantum] functions" (italics mine).[24]) – in order to be distinguished from every other physical thing/ … . Time for "now of der Augenblick", with reference in the affect *surprise-startle* to "aggregate" with the operative moment of *expectation?*

For now, we have the one-year-old who does not know very much about how s/he "thinks". What we know is that the one-year-old is equipped with an operative capability that engages in "one and the same physical thing/…" as "in the constant moment", "one and the same physical thing/…" among others inserted in "when/synchronous/Augenblick". To that may be added what Bejarano proposes for the one-year-old, namely that s/he has "distinguished number 1 in the series of numbers of 1, 2, 3 …" (order). To me this ought to be possible only with an operative functioning that considers "moment plus when/Augenblick" in time. This I'm prepared to suggest in terms of expectation on the way to becoming *intention*. It is intention that is the incentive of finger pointing, in its turn anchored in gaze pointing, with the aid of the eye's seven distinct muscles for its movements (the only sense that is "muscle-governed"). In finger pointing as an operative expression, we may see that it does not include (i) the contingency affect (surprise/interest) → gaze → relationship eye-to-eye; (ii) "moment plus now/Augenblick" as a feature of time extending to the next of "moment plus now/Augenblick"; (iii) expectation converted into "intention". What we experience from the number order of the "triad" in the form of (i) the one-year-old child's pointing to the physical thing/…; (ii) "other/caregiver" who answers to the pointing (for instance by taking the physical thing/… and handing it to the one-year-old, by giving it a name

24 R. Crease and A. Scharff Goldhaber, 2014, p. 120.

etc.); (iii) the one-year-old who receives the physical thing/... repeats/imitates the vocalization of it etc. What we do not know anything about is what expectation/intention broods on regarding the *unexpected* physical thing/... . To my knowledge, Aristotle never asked the question what alternatives there may be (of affect/internal presentation) that can exist (beyond efficient cause) within final causality. What we may assume is that with counting abilities, ultimately finger pointing, the brain has developed a new centre towards the periphery, and vice versa – something that is valid for the appearance of all new emergences during ontogenesis.

How was it then with Aristotle and final causality? For him the world and the cosmos were defined by divinities with objectives – the inheritance from Plato. Today we know that (final/formal) causality, ultimately anchored in expectation a priori and a posteriori with an extension to billions of galaxies in the Universe, is, if not immeasurable, at least undefined. Expectation does not only focus on the present (physical thing/...). There is always an extension to the absent (physical thing/...), as well. What Aristotle did not consider enough was that the human being, the development of society and its technology serving as the platform, is indeed equipped with objectives – despite all the constraints there are to pass them – without ends.

Is there anything else that is so attractive, literally speaking, (to the gaze) as finger pointing?

Two years old: from something to nothing to something

At this moment/now our one-year-old has become two years old. At the age of two years the infant has acquired the skill to sensorimotorically imitate a sequence of "moments plus now/Augenblick" – a sensorimotor event – and to "punctuate" (Wilden) it by another sequence of "moments plus now/Augenblick", another sensorimotor event. This also includes that the child may keep the event in focus without the immediate need to repeat (imitate) it.

In my fourth book in Swedish I give an example of the two-year-old child who understands that the overturned cup, out of two cups filled with make-believe water from an empty jug, is the "empty" one. What we know today about the two-year-old is that s/he is equipped with affects that are starting to merge into emotional script patterns, has acquired the ability to form recognitive internal presentations; that the ability for evocative internal presentations is under development (see Figure 3.1 under the heading "The developmental line of affect/internal presentation", to which I will return); moreover, has access to language with the capacity to create at least short juxtapositions of words and sentences. The question I ask about this is what is required *operatively* for the two-year-old to understand the event described above, in which *no* water is spilled and how this is possible when the cup from the beginning is empty, thus, the whole event is "make-believe".

Expectation in the human has reached truly staggering heights, beyond comparison with any other primate. Already in expectation a posteriori there ought to be an absolute difference between the newborn and other primate children. I suggest that at least a part of the explanation why the "no water" in the cup may be "poured out" for the two-year-old, can be found in expectation with the immediate leap from the present physical thing/... with a moment of elongation to the non-present/absent physical thing/... . If we are to follow Teresa Bejarano, the "no water" is an "empty profile to be filled up" – but the question one may ask is what this "empty profile to be filled up" is "filled up" with, by the two-year-old. In order to understand anything about how this may come about, I take as a vantage point the time-reversed expectation a posteriori in relation to the expectation a priori. In line with Bejarano's opinion, outlined in Chapter 1, expectation a posteriori is included as an element in the infant's mimicking of "the caregiver's/other person's behavioural and kinaesthetic expressions". As is made clear in my fourth book in Swedish, other primates (in the wild) lack access to the ability to mimic connected to expectation a posteriori. The question then is: What may the difference be between the human infant and the chimpanzee infant if expectation a posteriori beyond the present caregiver/other/other thing/object is put against the non-present caregiver/other/other thing/object?

A unique situation, as is stated at the beginning of this chapter, is that the infant, in contrast to other primate infants, is surrounded by a perpetual "buzz" of spoken language. Another unique situation which is different is that the caregiver/other is sometimes equipped with *physical things/ manufactured objects/artefacts* when s/he is present within the range of the child's visual perceptual field, and *sometimes not.* These physical things/ manufactured objects/artefacts may, consequently, be present together with the caregiver/other in a certain moment – in the next moment they may *not* be present, thus absent. My suggestion now is that mimicking linked to expectation *a posteriori* has its origin in the *Homo* line's capacity for inventions and that the gradually increased use of *physical things/ manufactured objects/artefacts, has successively converted into technologies in a never-ending stream.*

But let's start from the beginning. Alison Gopnik describes how it ought to be when the caregiver/other within the range of the infant's visual field picks up a ball, for instance, which becomes a "time event" for the infant, as long as the ball is kept in the arms it constitutes "no time event at all" and when the ball is put back on the floor it becomes the next "time event" for the infant.[25] My primary interpretation of this course of events is that during the first year of life visual perception is not yet diachronous, the utmost characteristic of the grown-up's visual perception. Moreover, the operatively different events – picking up the ball, and putting it down – are joined as movements with different goals, final causal movements, that still, for the neonate, temporally take place in different moments – thus, without considering time as a condition. To conclude, physical things/ manufactured objects/artefacts become, already for the newborn, elements in expectation a posteriori, either as present or as *not* present, consequently absent. This ought to be a circumstance that successively distinguishes the physical things/manufactured objects/artefacts – "otherness" – from the caregiver/other as present, with or *without* these. The physical things/ manufactured objects/artefacts thus get, via expectation a posteriori, an early boost as inwardly formatted, later with features of time – "moments

25 Alison Gopnik, 2009/2010.

plus now/Augenblick" – in the human, beyond comparison with any other living creatures.

As a reference and precondition, we go back to Jonathan at the time of the celebration of his first birthday party. Among the presents there was a box with a Lego toy train set. Somebody picks out the pieces in front of Jonathan's eyes and puts them on the sofa he is standing at, and holding on to. Jonathan pushes the pieces, one by one, towards the floor. Finally only the open box is left. Jonathan picks up the box, drops it on the floor and settles there himself. I observe how he sticks his hand into the emptied/empty box and obviously searches with his hand for "something", but finds "nothing" – and without the least look of regret, or frown of trouble. As if his searching was disconnected from the affective domain. Immediately after he leaves the box to its fate, in the next moment he crawls away in search of something new, anything that might come in his way.

It is through pretence play, which occupies the child between the ages of two and six or seven years, that nothing (the non-present) can become something (the present) – clearly without the operative function of the child objecting to it. Before I go any further in describing how this is possible for the operation, a remark on an additional capacity that the child needs to have access to, namely evocative internal presentation of the absent object (a capacity that I will return to later).

My suggestion when it comes to the two-year-old is that s/he can join, using Gopnik's word, a "time event", the physical thing/manufactured object/artefact as present together with the caregiver, followed by another "time event", the physical thing/manufactured object/artefact as *not* present, consequently absent (from caregiver/other) *in one and the same operation*. How can this be possible? What we know is that the two-year-old has access to a detailed sensorimotor mimicking of a sensorimotor action in the shape of a sequence of (sensorimotor) events, for instance, like the one described in the example above with the make-believe water – an ability that no chimpanzee has ever attained (and never will attain?). In the temporal perspective both the start and the end of the sequence are converted into "moments plus now/Augenblick". Because my proposal includes the suggestion that at the transition *from the sensorimotor*, on the border of becoming *primary cognitive operation* a conversion takes place of the operation from momentary via "momentary plus now/Augenblick" into temporally asynchronous and spatially asymmetrical functioning. The winning ticket for the cognitive operation is that it rests on the original and given precondition, namely the caregiver/other with the physical thing/manufactured object/artefact in the hand (or in the mouth) in order to be used – or *without* (the absent) physical thing/manufactured object/artefact in the hand (or in the mouth) *keeping* the validity of being usable. Because when it comes to the thing/manufactured object/artefact, for the two-year-old, in Piaget's words, s/he shows signs of having reached the ability for "object permanency", namely that it is always usable as present, but likewise as *not* present, thus absent. Thus, there is no mystery in that make-believe water, not present, is as "usable" and "valid" as real water.

A small aside: Opposites, ultimately what I define above as "something" and "nothing", may emerge from one and the same unit, as the example collected from the world of language in Sigmund Freud's posthumously published text *An Outline of Psycho-Analysis* suggests:

> Similarly, contraries are not kept apart from each other but are treated as though they were identical, so that in the manifest dream any element may also stand for its contrary. Certain philologists have found that the same holds good in the oldest languages, and that contraries such as "strong-weak", "light-dark", "high-deep" were originally expressed by the same roots, until two different modifications of the primitive word distinguished the two meanings. Remains of this original double meaning seem to have survived even in such a highly-developed language as Latin in the use of words like *altus* ("high" and "deep") and *sacer* ("holy" and "accursed").[26]

To distinguish between "meaning/semantics" and "grammar/syntax" ought to have been beyond Freud's horizon. It was only with Noam Chomsky's work *Syntactic Structures* that linguistics and semiotics seriously took on – even with a bias towards – recursive syntax.[27]

26 Sigmund Freud, 1940/2010, p. 31.
27 Noam Chomsky, 1957.

Three years old: the foundation for the paradox of human life is established

First, a reminder of the difference between *contingency system* and *discrete combination system*. In the contingency system the elements deictically flow into each other in a historical theme-rheme-flow with a perpetual mutual influence on each other. In principle, the contingency system might be forced open top-down with the uncovering of the linked elements as a result. How this is possible, with my experience as a psychotherapist, I myself come close to. The tools in psychotherapy are *semantics* in language as an entry, theme-rheme bindings, the relation and the possibility of an eye-to-eye encounter in the "therapeutic room". In the best-case scenario hidden deictic links between feeling/emotion/affect and the history of the client's past relations as a plurality of theme-rheme bindings – added via a type of "false asynchronous/asymmetrical(?)" formations of constraint/warding off/interface – might be uncovered in order to, in the best of cases, be joined again, regulated in a new diachronous context. Is a prerequisite for this an achieved primary cognitive operation?

The features of a discrete combination system are in principle (from what we know today) temporarily indeterminable (an idea Einstein never could accept as valid for the quantum level). The quantum leap is still the most common metaphor for how the elements in the discrete combination system are related to each other. If we are to follow Allan Schore's elucidation of the right "emotional" and the left "cognitive" hemispherical development during the first twelve months (see Figure 3.2), at the age of three to five years there is a clear dominance of the development of the left hemisphere relative to the development of the right hemisphere.[28] How does this fit with Sigmund Freud's early discovery of the "Oedipus complex", the three-year-old child's entry into the "Oedipal phase", in the perspective of the child's development of an affectively dramatic period? Without doubt, it appears to be a paradox that the "cognitive hemisphere" in comparison to the "emotional hemisphere" is the dominant one when it comes to growth – at the same time as the three- to four-year-old child

28 Allan Schore, 2003; Alf Nilsson, 2013, p. 71.

is in a dramatic phase considering the development of both relationship/affect/directed motive and internal presentation (to which I will return).

A crucial element in the three- to four-year-old child is the changed awareness (visual perception) of the time hidden behind space. Time, as an intangible "abstraction", starts to be perceived separately from space. With that follows the knowledge about time, with an extension forwards as well as backwards. What we know, with reference to Alison Gopnik, is that the preschooler lacks the ability for *time orientation* backwards. The child, who may very well recall an event in the past, is totally disoriented concerning *when* – five minutes ago, yesterday, or a year ago – the event took place. Even if the preschooler knows that the arrow of time is directed both forwards and backwards, s/he perceives (his or her own) death as an event within the frames of the room; the notion of "inescapable" still means nothing to the preschooler. However, from that age on the child starts to get a hold of quantities. In play with other children, if the child has beads in his or her pouch to share, the child distributes the beads so that everyone gets an "equal amount" of them. Now the child also starts to count objects – first up to three, and then with age the number increases. The child is equipped substantially earlier to convert cardinal numbers into things in the outer world than to convert speech/language into the inner world. The child lacks the access to "inner speech/language in silence" until s/he reaches school age (Teresa Bejarano, Alison Gopnik).

From infancy, over childhood, to the high school child

It cannot be questioned: it is Jean Piaget's and the Geneva school's "genetic epistemology" that lay the foundation for modern psychology's view of the child's development from reflex-sensorimotor operation all the way to a functional formal operative cognitive being. As has already been made clear, Piaget, in his last vast empirical work, *Studies in Reflecting Abstraction*, went on from his former notion of "formal operative cognition" to the notion of "reflecting abstraction" in order to define our most advanced abstract thinking. As also appears initially, in my first book in Swedish I have already given a detailed account of Piaget's whole view of the development of "intelligence" or "operative cognition". I consider this approach to be common knowledge today, not only within the scientific world of psychology but for all active professionals within psychology. Finally, just a short quotation from Piaget that gives an indication of the difference between concrete and formal cognitive operations:

> Concrete operations [6–7 years of age], though they have reached the level of valid deductions, remain partly spatial to the extent that they range over material objects and not yet over hypotheses stated in propositional form [at first with the development of formal operations at 11–12 years of age].[29]

An attempt to simulate human operative capacity: the modern computer

I start by quoting Terrence Deacon: "I believe that we still don't really know what it [computer and network] is".[30] He also suggests that we know too little about the digital. Deacon believes himself to know one thing, though: "Without human intentionality, [digital] machine operation is just physical change."[31] In order to know more I take George Dyson's recently published book *Turing's Cathedral: The Origins of the Digital Universe* as my starting point.[32]

The year was 1936. That was when Alan Turing presented the article "On Computable Numbers, with an Application to the Entscheidungsproblem". In the article he formulated the still valid principle for producing the digital machines of today, all the way up to the Internet, iPhone and apps. But at that time the philosopher and mathematician Gottfried Wilhelm Leibniz (1672–1716) had already identified "binary coding as the key to a universal language ...".[33] With reference to simple algorithms Leibniz showed how the algorithms were "performing the basic functions of arithmetic as mechanically iterated operations on strings of zeros and ones. 'In Binary

29 Jean Piaget, 1977/2001, p. 300.
30 Terrence Deacon, 2012, p. 371.
31 Terrence Deacon, 2012, p. 496.
32 George Dyson, 2012.
33 George Dyson, 2012, p. 104.

Arithmetic there are only two signs, 0 and 1, with which we can write all numbers.'"[34] What Leibniz revealed was that "words" that code for order are dealt with in the memory as if they were numbers. What has since been revealed is that it is numbers that *do* something, and not only mean something that make the construction of the digital machine possible. Through Alan Turing's framing of this into a mathematical principle, the problem has since, in many respects, been about materializing the digital computer with the aid of mathematics as language. In order for such a machine to work, an important element is loops that give *negative feedback* with "the signals from the goal are used to restrict outputs which would otherwise go beyond the goal".[35]

Is it possible from this quotation to try already at this point to indicate what separates the digital machine from the digital principle transformed in the human being? I will give it a try. Let's take as our vantage point Teresa Bejarano's idea of the child who is born with expectation a priori – expectation always with extension towards fulfilment. In the world of the human being, already in the neonate infant, it is given that expectation is not always fulfilled – which implies that expectation ought be able to "take a beating", but absolutely not to be erased; expectation is able to handle "noise", because expectation has a side facing what is *not*, or *nothing*. As I understand it, this cannot be the case for the computer, to which every form of noise is detrimental. Could this have something to do with the handling of time?

Alan Turing's brilliant idea to get the "universal Turing machine" to work was to create a "square" for space separated from a "square" for time – to handle time as a "discrete" unit. The fundament for the digital machine rests on the replacement of time with a "a sequence of events" (which always take place in space) – concurrently the predicted event is replaced by "structure". What Turing also understood is that a "redundancy in addition to reduction of noise" is additionally required. In life on earth the human is always circumscribed to perceive time as continuous,

34 George Dyson, 2012, pp. 104–105.
35 George Dyson, 2012, p. 114.

or diachronous, through visual perception. Is this, down to the smallest detail, the whole truth?

We have all heard about the computer's smallest information unit being the "bit". With that as my starting point I will make an attempt, with aid of a quotation from George Dyson, to convey what is at stake for the digital computer:

> The fundamental, indivisible unit of information is the bit. The fundamental, indivisible unit of digital computation is the transformation of a bit between its two possible forms of existence: as structure (memory) or as sequence (code). This is what a Turing machine does when reading a mark (or the absence of a mark) on a square of tape, changing its state of mind accordingly, and making (or erasing) a mark somewhere else. To do this at electronic speed requires a binary element that can preserve a given state over time, until, in response to an electronic pulse or some other form of stimulus, it either changes or communicates that state. "Most of the essential elements or 'Cells' in the machine are of binary, or 'on-off' nature."[36]

It thus boils down to a "bistable circuit" which stays in the condition that you (the digital computer) put it in. Do we know anything about this state of time? I quote Dyson: "Distinction between a dot (0) and a dash (1) has to be determined in 0.7 microseconds, by 'inspecting' the character of the faint secondary pulse generated when a given location is 'interrogated' by the electron beam."[37]

Dyson concludes:

> Turing's machine [...] consisted of a black box (as simple as a typewriter or as complicated as a human being) able to read and write a final alphabet of symbols to and from a finite but unbounded length of paper tape – and capable of changing its own "m-configuration," or "state of mind".[38]

Complicated as a person? No, hardly. But this does not imply that Dyson thinks that it is impossible to construct a computer with, as he puts it, a "human intelligence". As I understand it, he hereby states something that

36 George Dyson, 2012, p. 124.
37 George Dyson, 2012, figure page 13 between pp. 136–137.
38 George Dyson, 2012, p. 247.

I define in this book as "operative cognition". With a nod to how Turing perceived the potential of the computer, I quote what Dyson claims regarding the relation between the digital computer and *Homo psychicus* in the whole of his complex plurality:

> The paradox of artificial intelligence is that any system simple enough to be understandable is not complicated enough to behave intelligently, and any system complicated enough to behave intelligently is not simple enough to understand. The path to artificial intelligence, suggested Turing, is to construct a machine with the *curiosity* [read "the positive affect of *interest-excitement*"] of *a child, and let intelligence evolve.* (italics and brackets mine)[39]

Along with Alan Turing, John von Neumann, active at Princeton University, ought to be the person who has been the most influential for the development of the digital universe and cyberspace. Von Neumann was the first to urge for the construction of the digital machine. At the same time he realized what such a machine may achieve – and where the boundaries are in relation to the living organism. Dyson gives a quotation from von Neumann:

> I would suspect, that *a truly efficient and economical organism is a combination of the "digital" and "analogy principle"*. [...] The "analogy" procedure loses precision, and thereby endangers significance rather fast ... hence the "analogy" method can probably not be used by itself – "digital" restandardizations will from time to time have to be interposed. (italics mine)[40]

Converted into the human being, you as a reader understand that I recognize von Neumann to be both ahead of his time and remarkably insightful with regard to what a computer may have as its "parallel", and at the same time lack the preconditions for, in relation to everything that constitutes the fundament for what a human being is.

A modern author like Terrence Deacon, working along the same lines as von Neumann, is worth quoting in this context:

39 George Dyson, 2012, p. 263.
40 George Dyson, 2012, p. 234.

> Alan Turing showed how, in principle, every valid mathematical operation that could be precisely defined and carried out in a finite number of steps could also be modelled by the actions of a machine. This is the essence of computing. A "Turing machine" is an abstraction that effectively models the manipulation of symbols in an inferential process [...] as the action of a machine. Turing's "universal machine" included a physical recording medium [...]; a physical reading-writing device so that symbol marks could be written on, read from, or erased from the medium; and a mechanism that could control where on the medium this action would next take place [...]. [...] To simplify a bit, the problem lies in the very assumption that syntax and semantics, logic and representation [read "internal presentation"], are independent of one another. *A [Turing machine] is a syntax without semantics and logic without representation.* (brackets and italics mine)[41]

I conclude this section by presenting my own thoughts about why it will never be possible to construct a human computer. In short: the computer is a project of the twofold, founded on the binary speech system as "language", with the speech system "doing" something. Sometime, perhaps, it will be possible to create a computer that fully simulates the human twofold converted into the cognitive operation's either-or. But the problem is about the utmost human, which is grounded in the congenitally and biologically given of the onefold, more or less the affects. I cannot see how the onefold of these affects can be transformed into the digital twofold without damaging the actual essence of the affects.

The digital and the human central nervous system (CNS)

Is there anything in the human central nervous system (CNS) that can be described in terms of "the principle for the digital code"? I turn to Anthony Wilden's classical work *System and Structure* in order to find ideas. The point of departure is Wilden's distinction between "binary differences" and "binary oppositions". "Binary differences refer to discontinuities, thus to differences between differences [...], outermost 'the specific difference'." The notion associated with "the specific difference" is "absolute" in the sense that

41 Terrence Deacon, 2011, pp. 90–92.

the interface between "the infinite number of differences is determined, [...] 'the *distinct* difference'".[42] This ought to be valid for "brute facts", as already mentioned, today only a concern to natural sciences. "Binary opposition" is a term coined by the structural research tradition. The pioneer was the anthropologist Claude Lévi-Strauss.[43] As I have already touched upon, the problem is about, as an organizational principle, putting contrasting pairs/dichotomies/antagonists against each other. What is required in this connection is that the contrasts, or oppositions, which are put up against each other in pairs, mutually exclude each other. The contrasts are normally complementary, but they may also be asymmetrical. This is the case when it comes to the bivalent, or the Aristotelian logic of A against not-A. The problem for the sciences in this regard is to create more or less perfect dichotomous *notions*. At the extreme, I suggest, it hangs together with the issue of the research strategy being directed towards "institutional facts" that are never given once and for all in an absolute sense. As is made clear above, both John Searle and Chris Knight suggest that a theory that includes both "brute" and "institutional facts", would imply a "Theory of Everything". It is what Terrence Deacon, with his creation or "road map", also aims at (Chapter 2).

Let us go back to the question whether the digital may have any impact in some way, somewhere in CNS. Wilden puts this in the following way:

> [T]he central nervous system involves neurons which receive quanta or packages of information via the axons and through the connecting synapses. Upon arrival at the synapses on the body of the neuron, these quanta are said to be summated, the result of which is either the firing or the inhibition of the firing of the neuron. That is to say, at the moment of "summation" (the process does not in fact seem to involve sums), the neuron either fires or does not fire. Thus the neurons may be said to operate digitally, but the synapse and axon which connect them appear to be complex analog devices.[44]

42 Anthony Wilden, 1972/1977, p. 159.
43 Claude Lévi-Strauss, 1962/1984.
44 Anthony Wilden, 1972/1975, pp. 156–157.

Let me add to the quotation from Terrence Deacon's "road map": What Wilden expresses as "inhibition of the firing", may in my understanding be linked to what Deacon defines as "constraint".

Today we know that the neurons in the brain work both electrically, with axons and dendrites, and chemically, through transmitter transferences (dopamine, acetylcholine, serotonin, etc.). A nerve impulse, or action potential emerges from a state of rest through discharge in the form of a depolarizing–repolarizing. Normally, the nerve cell has a negative potential in comparison to the environment of positive potential. The different potentials hang together with a high concentration of calcium ions intracellularly, and sodium ions extracellularly.

A modern brain researcher like Jean-Pierre Changeux, with the knowledge of how neurons communicate as a basis, has studied how changes in larger structures in the brain may influence the interaction between neurons. Such larger structures, in the form of "specially distributed networks", work according to the principle of "neural diversification". Changeux describes, among other things, how existing neural networks give rise to the reflex movements in the newborn infant. Such movements, for instance grasping an object with "operative efficacy" (my wording), "are selected" and reinforced successively as their synaptic liaisons become stronger.[45]

In order to find *the* most fundamental explanation why the digital became a supplement in life, let us go back 500 million years in time. We may attribute to the sea the boundary for the development of certain species of the phylum Chordata into vertebrates. And the all the more crucial transformation constituted in the ossification of the spinal cord resulting in the development of the spinal column, and eventually in bone formation. With this, jaw formation followed in fish. Instantly, fish could from within, and via the opening of the mouth, catch food from the outside (in the sea) – and give it room for digestion. Thereby, the instant became prominent as "discontinuous" or "digital". When necessary prey became larger in size and range, the boundary between bite and chewing reflexes was blurred. Bone formation has become an *inner* hard kernel that distinguishes

45 Jean-Pierre Changeux, 2002/2009.

it from an *outer* hard shell. Bone formation is the prerequisite for acting. It constitutes the diametrical difference in *moment* – within cosmology the *momentum* for the original start of the Universe – which is the origin of the executive in its whole. But without biphasic regulation the ossification would not have made it other than as a skeleton.

With these examples, which may be understood as CNS, with its linkage to ossification, except for the analogue functioning – the most common standpoint – in certain respects, has elements which can be deciphered according to the code of the digital principle.

The developmental line of affect/internal presentation

Silvan Tomkins was the first to follow up Charles Darwin's detailed observation in *The Expression of the Emotions in Man and Animals*, namely, that there are a number of "innate emotions". I have distinguished, in Darwin's work, at least pain, distress, joy, pride, anger, disgust, surprise, fear, shame, each with synonyms, as universal, and they can be distinguished in both humans and animals, most explicitly in primates. With a unique tracking sense Tomkins traced the human fundamental and thus the given innate and biological affects.[46] If we follow Tomkins, the affects are nine in number: two positive, *interest-excitement* and *enjoyment-joy*; six negative, *fear-terror*,[47] *anger-rage* and *distress-anguish*, "origin affects"; in addition *dissmell, disgust* and *shame-humiliation*, the *auxiliary* affects. The ninth

46 In his first two books, published in 1962 and 1963 respectively, the number was 8 – only to be increased to nine in the posthumously published book in 1991, the ninth given name of the term *dissmell*, coined by Tomkins.

47 About "fear" Charles Darwin, 1872/2009, p. 267, percipiently wrote: "The word 'fear' seems to be arrived from what is sudden and dangerous; [...] I use the word 'terror' for extreme fear; [...] Fear is often preceded by astonishment" [read "surprise"] (brackets mine).

and last is the "neutral" affect *surprise-startle* that works as a "refreshing" or a "recalibrating" affect.

With a specific 4D-ultrasound scanner Nadja Reissland *et al.* were able to register two distinct facial expressions in a foetus of thirty-five weeks, "cry-face gestalt" and "laughter-face gestalt". This suggests that the child at the time of birth is well prepared to convert affects into facial expressions. In Figure 3.2 the two expressions of affect are obvious.

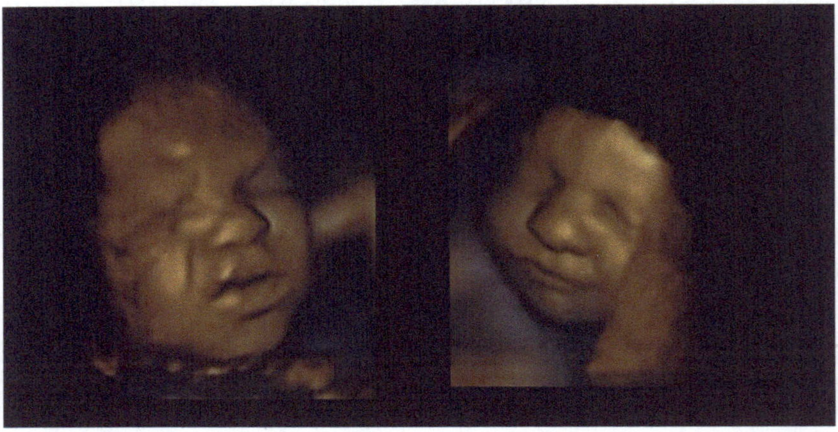

Figure 3.2. Cry-face gestalt and laughter-face gestalt in a foetus, pregnancy week 35. From N. Reissland et al.[48] Permission for use granted by Dr Nadja Reissland.

Early on I appropriated Tomkins's thesis that the affects are the human "primary motivation system" – but also the neonate child's fundamental "value system" for creating a first "meaning" about the world and himself. Affects in the human world function as *analogue stimulus and response amplifiers*: positive affects making good things better; negative affects that make bad things worse. Tomkins's notions for how affects work, among other things, are thus "amplify" and "amplification".

What is valid from the beginning of life is that affects and relations are so tightly conflated that I have joined them in a generic *contingency system*: Someone in Relationship to Someone Else, in short S-R-SE.

48 N. Reissland, B. Francis, J. Mason and K. Lincoln, 2011.

Specific properties for the affects, besides being *analogue*, are that they are *independent*, *general*, *self-amplifying*, *contagious* and *abstract*. The latter imply that the affects inform that *something* takes place – beneficial if the affect triggered is positive, detrimental if the affect that is triggered is negative. But affects never tell us *what* is happening, or *how* it should be dealt with. The human being has other structural/functional systems – the body as a vegetative system, acts of handling, perception, evocative images, cognition and language – to tackle *what* and *how*. Affects are also, besides the fact that they are the most indigenous ontogenetically (together with the reflexes), the most *contagious* we have in communication with each other.

Surprise, fear and interest are *gradient or stimulation increase affects*, surprise having the largest steep rise and interest the smallest (in between one finds the rise of fear). Enjoyment is a *stimulation decrease affect*. Anger and distress are *stimulation level affects*, we have anger in order to endure the fight until victory is won, and distress, a derivate of anger, we have in order to lick our wounds after surviving the lost fight, and to let the "pain" wear off; the latter Tomkins regard as a "semi-drive" (space- and time-specific, see below) in order to be able to return to life. According to Tomkins, interest and shame are the "disregarded" affects. Shame is the negative reverse side of "interest" – an incomplete rupture of the interest triggers the shame. It is interest and shame that are the most important affective contributions to the child's eventual development of relations, internal presentations and socialization. Considered in the evolutionary perspective, it is the affects that have contributed the most to the development of *Homo* into *Homo psychicus*.

Tomkins differentiates between *affects*, *emotions* and *feelings*. This he does within an ontogenetic framework. Affects are, as already mentioned, biologically given. If *affects* are biology, *emotions* are biography, thus connected to the experiences the human being has from the beginning to the end of life. Emotions are inner pattern formations (Tomkins's term is "script") – and as such are governed by the "principle of resemblance" or the "analogue principle"). The watchword is "recognition" – but "recognition" in this context always takes place on a psychological "unconscious" level. It is only at the point of transition to adolescence that affects come in the shape of conscious *feelings*. With reference in the affects interest and shame it leads up to the emotional spectrum: self-transcendence/alienation (commonly described as "shame") – responsibility – guilt.

Affects as onefold

The precondition is the prematurely born, helpless, vulnerable and dependent child. For several months s/he is 100 per cent dependent on the motherly world. In order to communicate with that world the child has access to the innate affects – which are in their turn answered and tuned in to by the mothering world and its affects, and their further development into emotions and eventually into feelings. It is given, as already mentioned, that the affects function as analogue stimulus and response amplifiers: a good response leads to positive affects in the child, which makes the child sense better: a bad response leads to negative affects in the child, which makes the child sense worse. During the first phase in life the child "knows" nothing about "good/better" or "bad/worse". The child has not yet acquired a foundation for proper boundaries between the infant and the mothering world. For the child, this "homogeneity" is "taken for granted".

Within the framework of onefold the child is embraced by the motherly world at the same time as this world is a part of the child. We do not know exactly how the affects tie up with each other and form possible "affect scripts" each time the child gains some new experiences. But my suggestion is that the early affects, together with acquired experiences, form a sort of basic "value deposit" within the child. Because, as already mentioned, the child is embraced by the motherly world at the same time as this world is contained within the child.

Together with the analogue principle, the world of the child is regulated by the *synchronous* principle: Everything in a synchronous/now/Augenblick where the triggering affect decides the "value" and the "meaning". The child does not have access to either "otherness" or "the unfamiliar". I also propose that the basic amental value deposit remains and accompanies the child throughout life. Now we know, considered from the outside that it is in the *relation* to the motherly world that the "meaning" of the value deposit is decided. What we also know is that this meaning may be biased towards abundant positive affects at the expense of negative affects, or vice versa. Moreover, the centre of gravity may be skewed in its extension along a continuum from one pole to the other. What we do not know, however, is whether we may as grown-ups in a new relation, for instance, in a relation to a psychotherapist, go back the whole way to reach this value deposit in order to influence its "originally formed meaning".

Can we speak about a "developmental route" from the beginning and throughout the first year of the child? And that it serves the child in the process of acquiring attachment skills, involving: (i) gaining a foothold in the (motherly) world, (ii) gaining experience of an integrated body, (iii) from the vantage point of the more or less established foothold in the world and the experience of having a body more or less integrated, being able to separate what is directed towards "me-ness", from what is otherwise in direction towards "otherness"? According my proposal, as mentioned earlier, the hallmark for the neonate child is "onefold", where the thin membrane between the child and the world has ectoderm (disposition for skin, brain, central nervous system and the eye lens) as its foundation. The skin as a sensory organ is produced during pregnancy, before other sensory organs. For the neonate child there is on the one side of the membrane the world as fragments, on the other side there is the body as fragments. In the creases of the skin organs for tactile sense, light, sensation, contact, pressure and pain are seated, and so is the thermal organ for heat/cold. If we are to follow Didier Anzieu, the skin of the neonate child as a border is extra porous and permeable. This ought to make the borders between touching and being touched, between outer and inner borders, floating. What I denote as "onefold" Anzieu defines as "cenesthesia", a general sensation of existence in the world without borders between the outer and the inner world. To this we may add Tomkins's re-evaluation in 1991 of the role of the muscles in connection with the affects – the muscles are there for the action – instead he wanted to underline the close link between the skin and the affects:

> [T]hree of the most compelling states to which the human being is vulnerable arise on the surface of the skin. Torture via skin stimulation [...]. Sexual seduction, again

> via skin stimulation, particularly of the genitals [...]. Finally, fatigue to the point of extreme sleepiness appears to be localized in the skin surrounding the eyes.[49]

In the beginning there is no impediment whatsoever for the child in the eye-to-eye encounter. "The taboo on shared inter-ocular experience" (Tomkins) is a consequence of the socialization process of the child. I am prepared to propose three elements in the eye-to-eye encounter that contribute to the successively growing taboo against gaze contact, namely: (i) to reduce the affect contagion in order to keep it in control, (ii) to limit the escalation of the affects, (iii) to restrict the risk of promiscuity. Most of us know that in the now/Augenblick of mutual passion all impediments against eye contact are torn down – the enamoured who does not let go of the adorable other's gaze.

My most comprehensive and detailed rendering of Tomkins's view of affects is to be found in my first book in Swedish (2005). But of course, the truly interested reader I direct to the source, all three books have the main titles: *Affect, Imagery, Consciousness.* I have been told that the texts in these books are impenetrable. I believe, as Darwin wrote in a letter to his cousin Francis Galton (known for his view of intelligence as being exclusively a matter of heredity and selection by "eugenics"): "I have always maintained that, [...], men did not differ much in intellect, only in zeal [read 'interest'] and hard work ..." (brackets mine).

The affects – from origin to recasting into psychical phenomena

Sarah Blaffer Hrdy, in two momentous works, entitled *Mother Nature* and *Mothers and Others,*[50] has taken on the task of trying to map the change that occurred, in what I denominate "the motherly world", when the mammals emerged:

49 Silvan Tomkins, 1991, p. 12. See also Didier Anzieu, 1985/2000, p. 45.
50 Sarah Blaffer Hrdy, 1999/2000, 2009/2011.

> This is when babies began to be born so helpless that mothers needed to be attuned to the smell, sounds, and slightest perturbations in the condition of vulnerable young that had to be kept both warm and fed [via body contact and sucking].[51]

Hrdy does not hesitate: with the mammals "care" ("emphatic glimmerings") emerged in the world. "Caring for" is the prerequisite for "attunement" that is required for everything that has to do with imprinting, binding and, in humans, attachment. Contained within care is the protection of the offspring from threat/danger/attack. This, require that the mammal mother – and the mammal father – are equipped with the affects fear and anger. For you as a reader, my description ought to contribute to the understanding of Tomkins's derivation of the affects fear, anger and distress from the "the origin affects" themselves.

A question you might be asking yourself: Were the mammals from the beginning without any traits of having what Tomkins classifies in the human being as positive affects? In the quotation above, from Hrdy, there are elements that suggest what it might have been like in the beginning: "attune to the smell ...". In previous books I have repeatedly stated that it is in the sensory modality smell that the affect *interest-excitement* has its roots. On the other pole, across the negative axle, *dissmell* has emerged. With the emergence of *Homo*, the auxiliary affect shame-humiliation appeared and acquired an important role in the socialization process of *Homo* on the way to becoming *Homo psychicus*. Tomkins's suggestion, which has become a conviction, is that the affect of shame closely followed the original affects fear, anger and distress (in addition to the auxiliary affect dissmell) – characterized as they are by instantaneous fear, merciless anger and lengthy distress.

If we continue to consider the majority of the mammals, it is in many ways the nosing and the sniffing out that leads to a sense of well-being – ultimately converted into the affect enjoyment-joy in the human being. Walter Freeman and his co-workers have dedicated an extensive amount of research work to mapping smell – chiefly neurobiological, and further out towards domains and functions for which smell constitutes the basic

51 Sarah Blaffer Hrdy, 2009/2011, p. 39.

fundament: "The nose was and is the first and final arbiter of what we ingest and of what we are afraid."[52] In the world of grown-up human being smell has seemingly been overshadowed by the other sensory modalities – not least by sight. Sometimes studies are presented that show that smells of different kind can be crucial for (albeit in the background) the decisions humans make in certain contexts.

As is clear from the above, Tomkins considers interest-excitement (along with shame-humiliation) to be the disregarded affect. My take on this is that we have been induced by the cognitive and phenomenological tradition to believe that everything, in the beginning, is about attention.

The positive affect enjoyment-joy, with the smile and laughter as the utmost expressions, is probably reserved only for the human being. It ought to have forerunners in disparate forms of expressions in other mammals. To give some examples: the purring of the cat is today considered to be one of these forerunners, as are the "song" of the gibbons and the primates' sound of laughter.

Above I have repeatedly defined fear, anger and distress within humans as "origin affects". The three negative auxiliary affects – dissmell, disgust, shame-humiliation – I leave aside, in order to come back to them.

The ninth affect Tomkins derived, surprise-startle, thus remains. There is an account for this affect in Tomkins's first book, published in 1962, the year after I started my studies in psychology. On the reading lists for the courses in neuropsychology the new mappings of the brain, in the Western tradition defined as "arousal reaction", in the Russian tradition – Yevgeny Sokolov being the most prominent researcher – as "orientation reaction".

In the exposition of surprise-startle Tomkins undertook the task of examining the relation between surprise as affect and arousal/orientation as reaction. Here I will insert a long quotation from Tomkins to serve as a background to my own continuing reasoning about the fundamental role of surprise as "the arousal affect of arousal affects" (Chapter 4):

> According to Sokolov, the orientation reaction appears with the onset of a stimulus and is later followed by adaptive responses. [...] [Sokolov] has distinguished phasic

52 Walter Freeman, 2000, p. 20.

> from tonic orientation reactions. The former is what we are calling surprise-startle reactions, the latter we would interpret to be one of several possible alternative affects, including interest, which might properly be called an "orientation" reaction but not necessarily excluding distress or shame, which are not orientation responses per se. [...] We would argue that there are involved in the orientation reactions three distinct mechanisms: the first two, surprise-startle and interest-excitement being affects, and the third essentially a reflex. An analogous situation holds for feeding reactions in infants. The hunger drive and the concurrent affect of excitement are aided by two ancillary reflexes; first the reflex of rooting, the movement of the cheeks upon stimulation which produces a rotation which is continued if the nipple has not been found so that the stimulation of the turned cheek by the breast helps the mouth of the infant find the nipple of the breast, at which time the second reflex, sucking, is activated. Neither the hunger drive nor the affect of excitement is sufficient to guarantee early feeding without these additional reflex mechanisms of rooting and sucking. [...] A similar situation obtains between the affects of startle and interest and the orientation reflexes. The shift in attention to the appropriate stimulus within the stimulus field, for example, in tracking a visible object by head and eye movements, or the shift from one sensory modality to another, e.g. pricking the ears and turning the head after looking intently at something, are reflexes which occur involuntarily upon the relevant specific cortical excitation.[53]

The quotation might be considered to be a perfect driveway to my own view of the relation between expectation and the affect surprise-startle (Chapter 4).

In the chapter on surprise-startle Tomkins conveys his view of the relation between surprise and interest, as well:

> Despite some interference between startle and interest their more general relationship is one of mutual facilitation in which growing interest may activate startle, which then activates interest.[54]

I interpret Tomkins's opinion about this relationship as follows: The triggering of interest (and fear) as gradient affects is always preceded by activated surprise. The fact that interest may trigger surprise anew, I explain in terms of interest appearing at two instances in a row on a timeline – let

53 Silvan Tomkins, 1962, pp. 502–503.
54 Silvan Tomkins, 1962, p. 507.

us define these as "Interest 1" and "Interest 2", respectively. As has been made clear above, "the affects never seek what belongs to them", but always assist and serve other functions and abilities the human being has access to. Here I will settle on suggesting – in order to come back to a richer and more detailed argumentation later on – that between "Interest 1", which evokes a thought, let us call it "Line of thought 1", and "Interest 2", a formation of interface has taken place (see below) – a necessity for the ability of surprise to trigger interest anew, thus "Interest 2", but this time linked to a new thought, "Line of thought 2". An immediate conclusion that one may come to, against the backdrop of this relation, is that there is no mystery to the fact that the human being may take the leap from one thought to a new thought – and in the service of innovation, via the human access to transcendence within many fields.

Up to approximately one year of age, the infant's world is in many regards fused with the mothering world. This world is primarily governed by the affects. During this phase the child is defenceless in regard to the crude unreserved power game that operates within her/him.

The supreme perceiver Charles Darwin comprehended the state of things early on:

> The movements of expression [read triggered by "affects"] in the face [...] serve as the first means of communication between the mother and her infant; she smiles approval, and thus encourages her child on the right path, or frowns disapproval. (brackets mine)[55]

Our knowledge of how the child relates to and communicates with the mothering world has increased dramatically during the last few decades in the wake of extensive studies, and results obtained by "modern infant research". Among the pioneers, Colwyn Trevarthen, Daniel Stern, Edward Tronick and the husband-and-wife team of Hanuš and Mechthild Papoušek may be underscored. In my first book in Swedish there is a detailed account of the research up to 2005. The focus for this research has been to map the attachment process of the child.

55 Charles Darwin, 1872/2009, p. 333.

We know now, moving from the view of the infant as a being a passive receptive creature, that the child, from birth, to a large extent participates actively in his or her process of attachment. During the first hours of life the infant is "more awake than ever". This is manifested, primarily, in the "widened eyes" that give the persons in the mothering world "an intense feeling of eye contact". In the surrounding world it is (the picture of) the human face that mostly contributes to the child's interest. Allan Schore describes it in the following way: "[D]uring [...] eye-to-eye transactions the infant's maturing right hemisphere is '[psycho]biologically attuned' [...] to the output of the mother's right hemisphere."[56]

The trigger of motherliness is more than anything the infant's gaze activity. Daniel Stern calls the eyes of the mother "supernormal stimuli".

At the age of four months the eyes of the caregivers "are more prominent parts of the body than the mouth". As is evident from the above, the child at the age of six months may apprehend that the caregiver's gaze is occupied by some third in the dyad. The child follows the caregiver's gaze. If this third gives rise to a specific affective expression in the caregiver's face, then this affective expression is reflected in the child's face. At the age of nine to ten months the duration of the mutual eye contact between child and caregiver becomes shorter. This ought to be connected to the fact that the "strong affect intensity becomes intolerable". Up to the age of one year, the child looks more often in the face of the caregiver when he or she expresses a positive affect than a negative affect. At this point the development of affect/internal presentation within the child has taken a dramatic leap forward; I will return to this subject under the next heading. In sum, one cannot deny that Tomkins were on the right track as early as in his first books, with his underlining of the importance of affect relations already in the child's very first instant, expressed in terms of "face-to-face, eye-to-eye". Ever since my first book in Swedish I have returned to a formulation by Tomkins that sums up in a nutshell the results observed in modern research on infants:

56 Allan Schore, 1994, p. 76.

> Man is primarily a voyeuristic animal not only because vision is his most informative sense but because the shared eye-to-eye interaction is his most intimate relationship possible between human beings.[57]

Another author who has formulated the importance of the gaze contact did so in the following fine way: "As adults we reserve this human interpersonal glue for our romantic relations, or else we would have ended up in perpetual romantic entanglements."[58]

One year of age: Sigmund Freud's fivefold "logic" as insufficient

As is clear in all of my five books in Swedish, I embrace Jean Piaget's thesis that the infant is born without internal presentations, thus as amental. In comparison with other mammals, primates included, there is a clear difference between how the newly born mammal infant and the human infant is equipped. As an example I describe above how chimpanzee and bonobo (pygmy chimpanzee) infants early on, from within themselves, grip and hold on to the fur of the mother during her tricky flight from branch to branch, in tree after tree. The apparent difference from what the newborn human child may (not) perform has its origin in the bipedalism of *Homo*, the adopted narrow pelvic structure of the woman giving birth and a pregnancy of only nine months. Richard Leakey suggests, against the backdrop of the adult human's rich and essentially different equipment when compared to that of the adult chimpanzee/bonobo, that the female would need a pregnancy of twenty-one months (but the pelvic structure of the human female would never allow giving birth to such a giant baby, see above). The extreme prematurity of the human infant needs to be handled and compensated with extreme patience and prolonged care (with a childhood that stretches until adolescence/puberty). This is part of the conditions behind the suggestion that the infant ought to be born with no internal presentations or as amental.

57 Silvan Tomkins, 1991, p. 9.
58 Hans Erik Frölander, 2005, p. 15.

A small aside: In the comparative research targeting chimpanzee/bonobo infants and human infants, the research has in many parts its focus on the problem of how the primates' infants can be given practice and be trained to perform in a way that looks like achievements performed by human infants. The big and absolutely crucial question is what the *difference* between them consists of, and how it emerged.

Today we know quite a lot about when the first embryonic internal presentational form was established, constancy, within the child and how it manifests itself, so that we may distinguish it from the outside. Already in the 1940s psychoanalysts such as René Spitz and John Bowlby described what they called alternately "eight months anxiety" and "stranger anxiety" – and that it is the stranger's face that gives rise to the reaction within the child. Here, very briefly: in previous books in Swedish I use, in line with Silvan Tomkins, the term "stranger shame" for this as a generic phenomenon. In fact, the phenomenon might manifest itself through a whole range of expressions: from intense cry, and as such with elements of distress, via a frown of trouble and a staring gaze to expressions of "reflection". This is meticulously outlined by infant and child development research (I have described this research in detailed in my two first books in Swedish).

There are several ways of showing the existence of this first embryonic form of internal presentation within the child (Piaget used different experimental approaches). I usually refer to stranger shame as being an "absolute criterion". Stranger shame appears when a face of a stranger pops up within the *visual field* of the child, while the well-known caregivers are *out of sight*. I cannot conceive of it in another way than that the child has access to some sort of inner psychic internal presentation of the well-known face(s) to which the present stranger's face is matched – which makes the stranger's face dissimilar to the face that is well-known to the child. Piaget speaks about it in terms of the child having acquired the ability to have a *recognition memory* – which he refers to as the transition of the child from amental to mental, or a psychic being. As is made clear in Figure 3.1, I use the term *recognitive internal presentation*. Thus far, I round up by subsuming the recognitive internal presentation into the child's first developed psychic system, defined by me as the System of Recognitive Internal Presentation, hereafter denominated by the acronym *Srip*. For *Srip* the key words are "recognition", "similarity", "synchronous functioning", and the system is in its totality regulated by the "analogue principle".

Srip has several features in common with what Sigmund Freud defines as the psychic system *Unconscious*, in English often referred to by the abbreviation *Unc*. I have described in detail how this may be in a previous book in Swedish.[59] Freud gives five criteria for the "logic" that makes him distinguish *Unc* as a unique psychic system: condensation, displacement, timelessness, absence of oppositions/contradictions, and no distinction between inner and outer (on timelessness as exception, see Chapter 6). I adopt these criteria as valid in connection with *Srip*. I will give a broader description, beyond the five criteria given for *Srip*, in the continuation of this section.

Out of sight, out of mind is an old expression whose origins as have been traced back at least to the sixteenth century. Today the expression has been brought to the fore in the psychological literature. It is connected to the contemporary distinction established between the capacity for "*recognition* of the present object" – "in sight" – and the capacity for "*evocation* of the non-present/absent object" – "out of sight". I will return to the latter.

We are thus staying on in *Srip*. When describing this system, I will take as my starting point the innate biopsychological affects and follow from there. In the evolutionary perspective, the rise of the affects coincided with emergence of the mammals. Closely connected with the emergence of affects are: (i) suckling, under the condition of direct body contact between child and mother; (ii) in Sarah Blaffer Hrdy's words, "care", which demanded more and more time and "patience" from the mother who had to satisfy the needs/requirements necessary for the child in order to grow up and become an adult individual, as more and more advanced species emerged.

Is there anything more helpless, vulnerable and dependent than a newly born human infant? The question has been repeated over and over in my previous books in Swedish, and at least once above. Why this insistency? It is due to my successive and elevated marvel over the fact that in the living world things are such that the most advanced and innovative, the human

59 Alf Nilsson, 2009.

being, has emerged via the rise of the most fragile and vulnerable in the living world, the newly born infant. What saves the infant is the access to affective bonding as means for communication. In this way, the relationship/affect has become a "value system" in the human world, far beyond the state of things for all other life forms on earth.

How are we to understand what Hrdy indiscriminately denominates as "care" in terms of affects as onefold? I have already mentioned the "nosing out" of the animals as the precursor of the affect interest of the human. As regards less advanced mammals, "care" requires "interest" in a form that I'm inclined to define as "caring for without nosing out as a precursor" – prying being fundamentally motivated by the drives. As I've touched upon earlier, protecting the infant from danger is included in the "caring" – the affect of fear as a motive of the mother – and attack – the affect of anger as motive of the mother – both affects are of concern for the mammal mother as well as for the mammal father. What was new and superordinate to everything within the mammal world was gaining access to the onefold "motive" of different domains in the world – everything prompted by the specific caring of the offspring, with demands for a "positive motive" without exception that "does not seek out itself". This "positive motive without exception" ought to be the source of the affects interest-excitement and enjoyment-joy, in the form that both affects are to be distinguished in the human being today. (Tomkins's suggestion is that interest and enjoyment as two separate affects, are reserved for the human being only.) My suggestion is thus that with the mammals the caring of the offspring became a concern/domain in itself, something that at the same time became the source of affects as the "primary motivation system" (Tomkins). Each affect started to develop in the direction of its "domain of survival *value*", and vice versa, each domain was tied to a specific affect in order to *analogously amplify* the orientation towards the separate domains in the world – affects as a kind of "thinking" for action (before *Homo* "thought" cognitively, s/he "thought" with the affects).

Could what I've just conveyed have anything to do with the emergence of *Srip*? My starting point is that there are no perceptions, thoughts, ideas etc. within the human being without underlying motives. This leads to the suggestion that, from the evolutionary bottom-up perspective, the first primary internal presentations became "motivating internal presentations" originating partly from domain-specific affects and partly from sense experiences of a certain type.

Against the backdrop of my previous arguments there are sense experiences/sensations linked to the operative functioning *expectation*, which immediately after coming into contact with a physical thing/ ... takes the (digital) step further towards the unexpected/not-expected/ next expected physical thing/ ... In this the foundation is laid for operative cognition/reflecting abstraction.

Thus, there are sense experiences linked to "configurations" – the first configuration (the iconic image of) is the human face – intimately associated with the domain-specific *affect*. In this the foundation for the Gestalt formation of perception is laid. Or, as an alternative, each sense experience that searches its way through the brain has two options along the journey: to follow the track that leads to the parts of the operative system formation, or to follow the other that leads to parts where the system formation for affects/internal presentations takes place (where one aspect of the long period of childhood development is to find and link the two system formations together, Chapter 4). These two alternatives for sense experiences may also be described thus: one as being in the direction of what physical things/... may be used for; the other towards the general affective meaning of human objects.

What is added to the emergence of *Srip* is, at the age of approximately eight months, the gaze and face of the stranger that is similar to the caregiver's face and gaze – but also contains elements of *dissimilarity*. The dissimilarity manifests itself in the shape of stranger shame, shame being the guardian at the border between interest and other negative affects, shame left out. But for an eight-month-old child with experiences of an abundance of negative affects at the expense of positive affects, shame is mingled with distress, thus, expressed as crying. But shame may also rush ahead the whole way towards anger, and/or fear, and/or dissmell, and/or disgust.

With *Srip* on its way towards more or less constancy, as a system formation, the first and the most fundamental *category formations* (as interface formations, see below) ought to have appeared: on the one hand the *well-known* (face) in the mothering world as a category; on the other the *strangeness* (face) as a category. This interface formation (interfaces are always formed via digitalization) takes place at the continuum: "abundance of positive affects/emotions at the expense of negative – abundance of negative affects/emotions at the expense of positive" as poles. How this interface and category formation appear, you as a reader ought to be able to perceive and understand, without doubt, that these have something to do with the child's (affective) experiences before and during the *Srip* formation. For the exposed child, the interface between the "well-known" and the "strangeness" ought to be characterized by both stiffness and vulnerability – the basis for what in clinical settings is denominated as warding off *splitting*. We may speak about warding off as a "superstructure" of what Terrence Deacon denominates "constraints". Parenthetically I would like to mention for the clinically interested reader that splitting tends to form clusters divided into two extremes: "idealization" and "devaluation" (as for instance, the borderline patient who oscillates between the two extremes projected sometimes onto "the well-known" converted into "me-ness", and sometimes on to the "strangeness" converted into "otherness").[60]

From *Srip* we may thus derive the first primary physical internal presentations as visual sense experiences recognized by affect derivatives and retained as *emotional* formations. In our adult perception, they are always present as governing emotive underpinnings, hidden by the evocative internal presentations. You may wonder how they influence what we denominate in ordinary speech as "perception". It ought to be in a situation that beyond the trigger surprise-startle gives rise to recognition – in a specific instant/Augenblick of a flickering space stimulus – immediately dealt with in *Srip* to be converted into System of Evocative Internal Presentation, *Seip*, as an evocative internal presentation. It might leave its mark on you in the shape of a vague déjà vu experience – sometimes it is easier to create a context around it, sometimes it is, for various (warding off) reasons, almost impossible. As soon as recognition is at hand, it lasts only as long as the key stimulus in the situation is still there in a hidden part of the visual field: Out of sight, out of mind. Thus, one may regard the eight-month-old

60 Alf Nilsson, 2013, pp. 144–146.

child's stranger reaction as the original emergence model for *Srip*'s whole structural and functional construction – so far only described by me from an outside and in perspective. How things eventually, during ontogenesis, may appear as from *within*-and-*in*, I shall return to later on.

Is the formation of *Srip* possible to understand in the evolutionary perspective? According to the principle of evolution "one takes what one has"; (unconscious) internal presentations in *Srip* may have developed as a conglomerate of affect derivatives and sense experiences with reference in basic (visual) configuration formations in the world, the human face being the primary source (face-to-face, eye-to-eye). A decisive factor ought to have been the *Homo* female's free hand carrying of the neonate child when breast-feeding.

Is it possible to distinguish a specific "targeted motive" with reference in *Srip*? As is made clear in my earlier books, "desire" and "Sehnsucht/yearning" (see Chapter 4) are targeted motives in the interface formation between *Srip* and *Seip* and between *Seip* and the System of Evocative Imaginative Internal Presentation (*Seiip*, Figure 3.1). Above I speak about internal presentations in *Srip* as "motivating internal presentations created by affect derivatives in a combination with visual sense experiences". The overarching categorization includes "the well-known" set up against the "the alien". The affect of surprise usually triggers *Srip* (but this is not always the case, the epileptic as well as the psychopath do not possess surprise). Thus, it is possible to define *Srip* as a psychic system formation that continuously throughout our lives puts us on track in situations that remind us of earlier experiences – but at the same time contains dissimilarities, against the backdrop of all situations experienced so far, thus, situations with strange elements/elements of newsworthiness.

Conditions for evocation: once more back to the beginning

As a human being I might take a step all the way back to where life began. The first living creature ought to have been blind, innocent, ignorant and without direction. At some point later on evolution/"natural selection" emerged. I do not hesitate to describe it as a totally ignorant and agnostic creature. But does this also apply to later eras when, at last, a first embryonic internal presentation had become an outcome, or precipitation, or emergence? In my view, the extension of the answer to the question is that at that point we have to abandon the description of "natural selection" as "innocent", "ignorant" and "agnostic". Embedded in the embryologic *morphodynamic internal memory, later on as internal presentation*, of whatever kind it might have been, an element of "content function" in the shape of the outer world with "meaning" in relation to the inner world, ought to have been present. Then we might argue for the emergence of a "predicative" element, exaptation, in evolution/"natural selection". What it had to deal with during this era, with morphodynamic emergence achieved, was the reconnection as an element in the exaptation in the form of re-presenting – still without us knowing anything at all about the teleodynamic addition, thus about the form/dynamics of this re-presenting. But it is worth being reminded in this connection about a deeper biologically founded relation: the mammal mother who "from-within-and-out" spontaneously gives birth to her infant – why not spontaneously "evoking" her infant?

Now I will take the giant leap forward in time to the era of the origin of *Homo*. It is important in this phase to distinguish between "internal presentation" and "sign" (in the sense of Peirce) in the shape of *symbols*, necessary for language. What we know about symbols is that they are arbitrary in relation to what they signify. Here I take Teresa Bejarano at her word: Symbol and language refer to socio-cultural evolution – because what would the symbols as arbitrary constructions, thus, as a construction of *difference*, mean in the biopsychological adaptation, or even in the re-presenting exaptation?

My answer is to stay put on the line of affect/internal presentation for similarities and dissimilarities – in fact, before differences started to summon up the operative development towards reflecting abstraction, and which led all the way to it. This is the one that is trying to drain macrocosm and microcosm of their content at the same time, from inside out and from outside in. Without digital operative cognition, we had failed with this "obligation" (note well, on the verge of removing all conditions for life on earth).

Silvan Tomkins does not hesitate in denominating affects as "analogue stimulus amplifiers". Within the frames of the analogue, one thing is similar to another thing, that is similar to third thing…. It is the analogue more or less. Because two phenomena can be valid: (i) a larger domain of overlaps, and thus a smaller domain without overlaps – which implies that both phenomena resemble each other; (ii) a smaller domain that overlaps, and a larger domain without overlaps – which implies *dissimilarity*. A sort of first step – or at least, an element of what ought to be the cause behind the leap forward towards operative cognition and its experimenting with differences; (iii) a similarly large domain that overlaps, as the one without overlaps – with emotion/recognitive internal presentation that oscillates in a state of *ambivalence* on the border of the paradox that is present when the operative cognition is fully drawn into the situation.

But let us go all the way back to the origin of affect as amplifier. At this point I follow Tomkins in his revision of the affect as *analogue amplifier of stimulus* towards, also including *amplification of response*, the whole way to perception, thought and the perception and the thought converted into language:

> The great German philosopher Immanuel Kant likened the human mind to a glass which imprinted its shape on whatever liquid was poured into the glass. Thus, space, time, causality, he thought, were constructions of the human mind, imposing the categories of pure reason upon the outside thing-in-itself, whose ultimate nature necessarily forever escaped us. I am suggesting that he neglected a major filtering mechanism, the innate affects, which necessarily color our every experience of the world, constituting not only special categorization of every experience but producing a unique set of categorical imperatives which amplify not only what precedes and activates each affect but which also amplify the further *responses* which are prompted by affects.[61]

In the mothering world the foundation is laid for the world view you have as an adult. Does this claim contain any truth value? My continued line of reasoning for this claim will stand in relief against two conditions: one of the conditions is Teresa Bejarano's derivation of three modes of gaze,

61 Silvan Tomkins, 1991, p. 18.

presented in my fifth book in Swedish; the other condition is Tomkins's revised opinion that includes response in the analogue affect amplification. The human being is the only animal with sclera, the white of an eye. Researchers within the cognitive tradition do largely agree that it is a structural/functional condition, "that makes it easier for others distinguishing eye direction".[62] Bejarano is not especially impressed by that interpretation:

> [A] showy white of the eyes is insufficient to guarantee genuine communication via eye direction. The ability to ascribe internal states to the eyes looking at me is also necessary for me to understand communication as such.[63]

My opinion is that Bejarano does not understand the whole depth of her own claim. Bejarano's project is to derive the syntax of language, with reference in cognition, deepest in expectation. But I am prepared without doubt to embrace her derivation of three modes of gaze.

In the world of mammals (I ignore the world of birds here), when a creature that fixes its eyes on another individual's eyes of the same species, then the gaze is perceived to have: (i) "immediate evaluation implication". Here I settle on two examples: one of the examples is that the staring on this level is the condition for the maintaining of the hierarchical order in the herd; the other example (which has been studied already) is valid for gorillas, where the gaze in an angry face of a male triggers fear in the newborn gorilla child; (ii) for the primates/great apes, especially observed in the chimpanzees, there is the possibility to release the ability to master the body in order to spot what another individual of the same species has within his or her field of vision (if it is not directly given). This is usually viewed as a homologue to what is denominated above as gaze direction detection. In the case of the chimpanzee, modus two contains, simultaneously (in one and the same gaze), modus one; (iii) the implication, valid only for human beings, is that from the very beginning we are stared at – the newborn child who perceives the caregiver's gaze, eye-to-eye, in the child's own gaze.

62 Michael Tomasello, 2009/2011, p. 74.

63 Teresa Bejarano, 2011, p. 59.

What is it then that the child apprehends in the staring gaze, eye-to-eye? With reference to Tomkins's theory of the analogue as amplification of – not only affect stimulus (the child presumably perceives that the caregiver is equipped with affects) but also amplification of affect response – the child finds for the first time a bridge to *communication* (and not to solely register affect eye-to-eye). This is reserved for the human being only. This communicative possibility I have previously described in terms of "the first language". With Anthony Wilden's distinction between "communication" and "information", information that requires language, as a starting point I will from now on speak about "affect communication" with the nine inborn affects as, metaphorically expressed, "letters", or more precisely put, as the basic elements in the affective construction of relationships, S-R-SE. But at the same time, I argue, in the same vein as Tomkins, that these basic elements later on can mobilize (via their development as emotions and feelings) "interesting" fields for knowledge (or "dangerous" fields etc. for the subject/group/society/earth in the biosphere ...).

A small digression as a reminder of that Emmanuel Lévinas constructed his whole ethical and moral philosophy with face-to-face, eye-to-eye as foundations for "the responsibility for the other" (see earlier books in Swedish). The premise for Lévinas is the human being eye-to-eye in relation to another human being – without closer scrutiny of *what* in the gaze relation or, *how* this relation/interaction shapes responsibility. Sometimes this is spoken of in terms of "ethical absolutism" or "ethical metaphysic". That shame functions "murderously" throughout for the subject in taking responsibility for the other, I regard as an insightful *psychological* addition to what is in other cases a philosophically abstract creation, top-down. Andrew Tallon, with reference to Silvan Tomkins among others, argues that "ethics as absolute" in Lévinas's thought may via "affect" open up Lévinas's closed ethical system. This may be possible through, among other things, "acceptance of the evolutionary perspective", in my wording, with ethics via affect viewed as bottom-up. With that "affect becomes a bridge between the social and the ethical". "Affect is the bridge from the social to the ethical, because it allows the advent of the other to be experienced in a passive moment, in a vulnerability to the other rather than in mastering

[...] characteristic of knowledge and will".[64] With Tomkins's revised theory of affect amplification, with *affect response* as an element in amplification, a totally new field is opened up and a deeper understanding of the line of affect/internal presentation as foundation for the emergence of other capacities during ontogenesis is made possible. Given, and embedded, as appears above, is the view of affect being communicative – affect as "contagious" beyond what we mostly use when communicating. Experimental psychology, as already pointed out, has shown that it is the facial *configuration* that is the first thing to arouse an interest in the neonate child. It is the amplifying affect response in the caregiver's face that paves the way for the child's re-presenting of some new outer object, and other inner objects. With language as a reference we may speak about amplification of stimulus as "theme", amplification of response as "rheme".

From what we know so far, within the amental infant it is not about "mentalizations" of the "outer new and other", but about an affect amplification in a system containing nine elements which can re-present (different) outer affect expressions. The principle is based on onefold analogue re-presentation, with the twofold expectation of the expected as well as of the unexpected/non-expected as a basic feature. Is this really the only and the whole truth? What I myself find worthwhile scrutinizing further is "the increasing affect of increasing affects", namely, the surprise-startle affect. But I will wait with that, as already mentioned, and return to the subject later on.

Back to the developmental line of affect/internal presentation. I have already described how the eight-month-old child re-presents affect derivatives and sensory experiences in the polarization of "the well-known" versus "the strange" as the foundation for *Srip* formation. In this connection, I also proposed that the implication of both poles was dependent on the experiences the child had already gained and later continued to gain during the formation of *Srip* – always in the ongoing affective/emotional communication, according to the rule "out of sight, out of mind". One may no longer dismiss the knowledge: deprivation, caused by absence or

64 Emmanuel Lévinas, 1961/1979; Andrew Tallon, 2007. Quotation pp. 667–669.

exaggerated intrusion by the caregivers in the mothering world, leaves its mark on the child, equipped from the beginning with affect communication. Extreme deficiency in the possibilities of having such a communication may manifest itself in the shape of *anaclitic depression*. For a detailed account of this I refer to my earlier books in Swedish.

General and independent aspects of affect

In order to find the fundament for the (in the visual field) evoked *absent* object, my opinion is that we should turn to the affect system, its evolutionary origin and characteristics, most of them described by Silvan Tomkins. An elaborate record of the affects' general as well as specific characteristics is to be found in my first book in Swedish about *Homo psychicus*.

When the affects emerged in the mammals, something very special occurred in the evolution:

> It is by virtue of its *generality of space and time* that [the affect system] can readily coassemble with and therefore impart its urgency and lend its power to memory, to perception, to thought, and to action no less than to the drives. (italics mine)[65]

My interpretation of this attribute of the affect is that it lends itself to "dependency", "mutual dependency" but, paradoxically, also to "independence". As we know already, this is hardly paradoxical: dependence and independence are "binary oppositions" – not "binary differences". Affect is fundamentally non-causal – which does not exclude that it might turn into a causal element that exercises a strong power. Affects may give structure to, in fact, everything a human being has access to as both a vegetative-somatic and a psychic creature – but they may also cause disorder and chaos. Why the idea of affects in the evolution?

Sarah Blaffer Hrdy, I assume, would refer to "care" as a necessary feature in the mammal world for the survival of the offspring; Dean Falk, perhaps, to the *Homo* mother's humming and simultaneous rocking of the

65 Silvan Tomkins, 1991, p. 72.

crying *Homo* child in order calm the child.[66] In this connection it might be worthwhile to stress Tomkins's view that the "affect is analogous to music": both are characterized by the common trait of being "abstract".

How is it possible for the affect to stretch from *dependency* over *mutual dependency* all the way to *independence* on the continuum? There is nothing in life (in the Universe?) that is absolutely independent. The mammal child's dependency (on the mother) I have commented on earlier in terms of it being a specifically stretched adhesive with origins in the emergence of the mammal. It implied also, by extension, the importance of the "specific relation", including *imprinting* (it would be far beyond the scope of this book to comment on the evolution of birds) and utmost the phenomenon of *attachment* that followed upon the emergence of *Homo*. In the perspective of survival, the mammal mother's "aim" was to see to that each infant would grow up to be as independent (and unique?) a creature as possible (in the service of reproduction, to be chosen for mating). To this process the stretchable affect *motive* could be of assistance (hunger can never assist thirst in its need for water, or the need of breathing for oxygen).

Now I take the big step from the emergence of *Homo* to the emergence of *Homo psychicus*. There is nothing as *dependent* as the infant. At the other end, there is nothing as *independent* as the grown-up human being who has gone, with perception, operative cognition and language as means, via the generality of affects (and independence), all the way to the Universe and its spatial extension: to galaxies millions of light years away; as well as all the way back in time, to the point where the Big Bang took place in the Universe, 13.7 billion years ago.

What is there to say about the affect *mutual dependence*? For an answer to that question I refer to Tomkins's revision of the theory of affect that not only amplifies stimulus – but also the response. It ought to be the source of the feedback processing with windings in the human brain – windings that are reserved solely for the human being. The inherent ability of the affect to also amplify response ought, eventually, to have led to the recognitive internal presentations re-presenting in the form of evocations

66 Sarah Blaffer Hrdy, 2009/2011; Dean Falk, 2009.

of our visual internal presentations of absent objects, at the extreme end. Before I continue, a brief interlude about what is usually termed "amodality", but which I prefer to denominate with the term *multimodality.* The availability of multimodality probably plays an important role in the act of re-presenting evocative internal visual presentations, ultimately converted into evocative images.

A digression into neurobiology

Neurobiologically, the conditions for multimodality are mapped in detail. Here I settle on referring to the surveys presented by Joseph LeDoux and Steven Pinker. The former takes the sense modalities as his starting point. Mostly they are triggered externally, but they may be triggered internally. The sense experiences normally have their reference in one or more of the sense modalities. From each of the cortical projection areas there are trajectories that converge in the rhinal cortical areas (*superior colliculus*) in the limbic system. There an integration takes place of the perceived experiences crossing the different senses before they are taken further to the hippocampus, in the same area where memory formation is founded. Thus the re-presenting in the form of internal presentations may be formed: "As a result, sight, sounds, and smells can be put together in the form of a global memory of a situation."[67]

Additionally, there is the inherent "equivalence" of the different senses. The foundation for the equivalence is that the areas including the thalamus and the projection areas for the different senses are wired so that they can "trace" each other and that one area, at least partly, may "take over" the function of another. Blind people take advantage of touch:

> using these prewired top-down connections to read Braille. They may be "imagining" the rows of dots as they feel them, much as a blindfolded person can imagine objects placed in his hand, though of course far more rapidly.[68]

67 Joseph LeDoux, 2002/2003, p. 104.
68 Steven Pinker, 2002, p. 95.

This way, the visual system in blind people may function analogously with how they function for seeing people, despite the absence of impressions from the eyes. Moreover, it is common knowledge within evolutionary and neurobiological research that eyes are formations of the skin and are composed of cells from the brain.

But we pause here for a while and scrutinize a little further the areas in the brain that LeDoux and Pinker single out as essential for the availability of multimodality. Moreover, the authors refer to structures in the limbic system, "the old brain", which is also to be found in other mammals. Against this backdrop my proposal is that a type of "primary multimodal" recognitive internal presentations not only are an asset for the human being, but also for other mammals. For examples of this I refer to my earlier books in Swedish. When LeDoux and Pinker speak about "representations" they do not distinguish recognitive internal presentations from evocative ones.

Three years old: evocative internal presentations: the small child's drawing as a criterion

After this digression into neurobiology, back to something I have described in detail in my earlier books in Swedish, namely, that sight is by far the most developed sense organ within the human being. As is made clear above, the re-presenting of phonetic sounds takes place evocatively at a substantially earlier point in ontogenesis than evocative visual internal presentations. This is often used as an argument that we "lack recollection-*images* before the age of three".[69] Is there any absolute criterion for the child at the age of around three years being able to gain access to the evocative visual internal presentations of the absent object? Research on children's ability to draw established early on – to my knowledge there is no contemporary research that has refuted these results – that children at the age of three years leave the phase of scribbling and create their first figurative drawing,

69 Michael Gazzaniga, 2008/2009.

the "cephalopods" (followed by "houses", see above).[70] The cephalopod, occurring in different cultures, is composed of a large head with marked eyes and mouth, the body marked only with a line, as are the arms/fingers and legs. As already mentioned, there is something about the "particularity" of the face – the face, the first stimulus configuration that the neonate child seeks with the gaze, and also, as we now see it, the first evoked visual internal presentation the child may elicit of an absent object. Is it possible to distinguish a "directed motive" behind the child's linked accomplishment to form that through operative sensorimotor activity, for the perception, figurative cephalopod at this age? The accomplishment, according to my proposal, has its reference in: (i) a specifically directed motive, *desire*, (ii) an operative sensorimotor and primary operative cognitive ability, (iii) an access to visual capacity to evoke the absent object that is available for continuous visual perception.

A brief interlude of visual perception and time: a proposal

Is it time that continuously follows visual perception in its footsteps? Or is it visual perception that continuously follows time in its footsteps? With visual perception follows, as nothing else in the human being, the principle of *diachronous* time. Thus, the directedness of an ordinary gaze during the course of running time. That does not exclude that the *synchronous/now/Augenblick* – Augen-blick! – may surprise the visual perception. Then, everything that takes place in the outer world is put into play with that which once upon a time has been stored in the barns of perception.

But things are not only about *synchronous recognition* when it comes to visual perception. Without access to evocative internal presentations, visual perception would not have been an asset for time as continuous and *diachronous*. My proposal is that *in visual perception there occurs a continuous evocation of now/Augenblick that makes time diachronous.* To this subject I will return in Chapter 6.

70 Lauretta Bender, 1952; Elisabeth Koppitz, 1968; Viktor Lowenfeld, 1947.

On the complex way of becoming a grown-up person, an important development takes place at the age of three years. I will wait with the derivation of the question of how this "block" ought to be put together at this age, to the next chapter. Maybe you wonder about the origin of visual images during ontogenesis? According to Figure 3.1 they would emerge, at least in a primary form, at around the age of six to seven years. Now my proposal is that they in their turn demand the emergence of additionally one specific motive, *Sehnsucht/yearning*. This too will be a subject for Chapter 4.

CHAPTER 4

Operations in relationships to affects/internal presentations: a teleodynamic investigation

> Cognitions coassembled with affects become hot and urgent. Affects coassembled with cognitions become better informed and smarter. The major distinction between the two halves is that between *amplification* by the motivational system and *transformation* by the cognitive system. [...] Amplification without transformation would be blind; transformation without amplification would be weak.[1]

Introduction

Ever since psychology, more than a century ago, became an academic discipline, a separation of the individual into, on the one hand, a being involved in motivation and affects, on the other hand, a being involved in cognition, has prevailed. In my view, there has neither been any research nor any schooling that has undertaken as its task to investigate how these two sides of human functioning could be dynamically integrated and provide our "sense of me-ness" as an adult person.

In my fifth book in Swedish I began an investigation of how to describe the dynamic relationships and interactions within humans between the operative line of development and the corresponding line of affects and internal presentations. This led to an extremely premature description. In this chapter my ambition is to present a more thorough and detailed

1 Silvan Tomkins, 1992, p. 7.

investigation of these conditions. In order to accomplish this, I would like to follow the two lines of development – evolutionary and ontogenetic – as near as possible, from the beginning and further on to the point of the important creation of the adult person's "sense of me-ness as a whole". As a researcher in clinical psychology (and as a practising psychotherapist) my experience is that such a sense could be disturbed during a person's epigenesis. My suggestion is that such disturbances would be a consequence of shortcomings and conflicts during epigenesis and the results could be psychopathologies of different kinds. What is thereby disturbed is a person's/patient's "sense of me-ness", resulting in a greater or lesser lack of sense of self-coherence. On the whole such disturbances follow the line of affects and internal presentations, in my opinion connected to what has been happening as regards the contingency principle of S-R-SE.

Surprise-startle, distress-anguish, expectation a priori, and expectation a posteriori at the beginning of life

The particularity of surprise-startle

In Tomkins's general description of the affects these amplify both stimulus and response. Can one find an exception to this rule? Concerning the positive and the negative affects one can hardly question that they are present as quality markers and also amplify the response. According to Silvan Tomkins, surprise-startle, as a gradient affect or stimulation increaser, is a conversion and a neutral affect. It is an ally of the other affects in that it triggers them phasically (Tomkins above) – but presumably does not contribute to "how" or "which of" the other affects that are triggered. Surprise has its own particularity. It hardly contributes to the amplification of the response. The affect supplies the response by indicating a direction: which affect has a certain positive direction, or which affect has a certain negative direction. By considering the element of affect in this way, I propose that surprise is not included in the amplification of the response (which does not exclude the idea that surprise-startle might appear in the shape of both emotion and feeling). Surprise is reactive and the advocate of the other affects, but in fact could also be defined to be the alarm bell of the whole organism. As such it also triggers the reflex where expectation is an inherent feature. Here I repeat what Silvan Tomkins has argued, which is outlined above (Chapter 3): "We would argue that there are involved in the orientation reactions three distinct mechanisms: the first two surprise-startle and interest-excitement being affects, and the third essentially a reflex." The connection to expectation a priori/reflex ought to be facilitated by the reactive element that gives a touch of the twofold's either/or to surprise. This I've so far only attributed to the specific features that may be derived from the operative line of development.

If one follows Teresa Bejarano, as I do, expectation is considered to be – both a priori and a posteriori – reactive. This means that expectation does not care whether the expectation is met/fulfilled or not. Thus, it is totally *instrumental* without any value-rating mechanisms. So far, I have made a point of distinguishing between two lines of development: one in principle adynamic, operative and functioning according to the digital principle; the other motivating, dynamic, and filled with affects, internal presentations and internal images, and functioning according to the analogue principle (language has its own particular position). The question one may ask oneself is whether there is anything that totally functions adynamically in the living world. The importance of expectation, in the form in which it can be observed in the human being, can hardly be overestimated in the development towards reflecting abstraction. At this early stage of the inquiry I take on the task of presenting an overarching proposal that will hopefully serve as both guiding and facilitating in the process of acquiring the account that will follow in the chapter: *Expectation, presumably, is from the very start supported by the affect surprise-startle, the latter presumed to be reactive, neutral and without influence over the response*. This has enabled the development of the innate twofold reflex the whole way towards the "neutral" and "objective" reflecting abstraction.

Viewed in this way, surprise opens up for the idea that it has the potential for penetrating the operative developmental line, as well as functioning as a fundamental bridge between the affect/internal presentation developmental line and the operative developmental line; a type of guarantor for *Homo psychicus* as an affectively relating being, as well as an abstract reflecting being. Simultaneously, this ought to be the basis for how a "sense of me-ness" could emerge as an experience within the adult person.

The susceptibility and flexibility of distress-anguish

But first of all, there is the birth scream/cry. About this Charles Darwin writes: "The earliest and almost sole expression seen during the first days of infancy, and the often exhibited, is that displayed during the act of screaming."[2] Is this triggered by an innate affect or reflex? Or by both? If so, triggered by surprise and expectation a priori? Or by distress-anguish? Converted into a terminology of anatomy and physiology, the human being's process of birth (in comparison to many other mammals) is long and complex: for the foetus, the onset of the breathing necessary for the supply of oxygen, in order to turn the bluish colour of the blood into reddish … . What I will focus on in the following is the affects surprise and distress, and also screaming/cry as expression.

A fundamental assumption in the book is that it is surprise that triggers all signs of life in the human being, from birth. Surprise is always the first step towards "recognition". This is already the case in the gestation period when the foetus apprehends the voice of the mother – but remains indifferent to the voice of the "other" until birth. The process of giving birth in the world of humans (in relation to the worlds of other mammals) is unique, accompanied with a *pain* (in the mother – but also in the child?) which is beyond everything else one may imagine in the living world (see Richard Leakey above). Here I will interrupt the account for a while, in order to place "pain", with the guidance of Silvan Tomkins, on the life line:

> The *drive* supplies vital information of where and when to do what. […] The *pain* mechanism is like the drive system in its place specificity. […] But unlike the drive system, the pain mechanism is time-general. […] Contrary to the imposed time

2 Charles Darwin, 1872/2009, p. 205.

> rhythms which are structural in the drives, the pain mechanisms is structurally time general. It imposes no time constraints on whether or how often it must be activated. [...] The *affect* mechanism, in contrast to both drive and pain mechanisms, is both space- and time-general (one of its specific characteristics). (italics and parentheses mine)[3]

One cannot understand the affect distress-anguish without considering the "semi-drive" (Tomkins) pain as an important foundation of human life. In the perspective of the emergence of *Homo* during evolution, distress occupied a particular position among the "origin affects" – fear, anger and distress. With labour you will bear children. Birth cry, and the sobbing (accompanied with tears from the age of three months) belong to the human only, as does the ability to suffer (from distress) in the face of his own death.

To my knowledge there is no proposal that has been fully embraced, about when birth screaming/cry could be observed for the first time in evolution. Dean Falk suggests that this might have occurred about 1.6 million years ago, at the time of the early *Homo erectus*. The emergence of the birth cry and the sobbing, in the form that might be distinguished in the human being, Falk continues, appeared in rapid succession. This coincided with one of the first innovations of tools – *baby slings*. The hypothesis behind all this, Falk formulates as follows: "Putting the baby down."[4] Regarding the birth cry, however, the commonsense opinion among evolutionary researchers is that it primarily had a communicative function, for the highly exposed and already utterly helpless *Homo erectus* infant, in the relation to the parents.

Does the birth cry have its deep roots in the inborn affect or in the reflex? My suggestion is that, at the moment of birth and during the first three months of life, there is a strong connection between, at least, the affect expression cry and the palmar grasp reflex. As early as the 1930s, research was able to show that an increased intensity of the cry of the child correlated with an increased strength in the grasp reflex, at least until the age of three months, when the grasp reflex starts to be dominated by the infant's

3 Silvan Tomkins, 1992, p. 71. For parentheses, see Alf Nilsson, 2005, pp. 101–102.
4 Dean Falk, 2009.

sensorimotor control. This relationship is basically regulated by our so often neglected "sixth sense", *vestibularis* – vestibularis is of importance to the multimodality like no other sense. Moreover, it is the sense that channels gravity from the beginning and that helps us keep our feet on the ground. And, as I have shown in earlier books, it balances the gaze through the inborn nystagmus reflex – and that is the reflex with an observable slow phase and a quick resetting phase, which might serve as a clear illustration of the rhythmic activity/biphasic cycle.[5]

Moreover, it is when the infant is three months old that the mother starts to distinguish cries that express "anger" from the ones that express "pain/distress" and "frustration" (with my addition of the affect "fear", see above and earlier books in Swedish).[6] To this, I would like to add my suggestion that the infant cannot express these three expressions of cry without surprise being present in the process of triggering the affects anger, distress and fear. Moreover, during the first three months of the infant, vestibularis is the interlinking system of the affect surprise and of the reflex expectation, the latter being totally under the control of the reflex mechanism. Above I have discussed the affects in terms of their being the child's "first expressions of communication". To that proposal, I now add the proposal that also the cry, backed up by the affect distress, is an element of the infant's "first expression of communication".

If interest and shame are the human affects par excellence, and surprise has its own particularity, then the affect of distress-anguish, with the human birth cry and sobbing, has become the *susceptible* and *flexible* affect. To a large degree it ought to be connected with the fact that the cry and the sobbing have their roots in pain, together with the other positive and negative affects, and are the child's "first expressions of communication". As such this guides us throughout our childhood up to adolescence/puberty – but

5 Dean Falk, 2009, pp. 31, 44, 185, 186. Alf Nilsson, 2005. Parenthetically, worth mentioning perhaps is that I started my research carrier with vestibularis in focus for my interest. It resulted in a doctorial thesis with the title *Motion stimulation effects viewed within an intrapersonal system of adaptation*, 1972.

6 Sarah Blaffer Hrdy, 2009/2011. For examples, see Alf Nilsson. 2009.

the cry and the sobbing are also an expression that may come in handy in specific situations in adulthood. I will come back to this subject later on.

Expectation a priori in relationship to expectation a posteriori: a suggestion

We already know, according to Teresa Bejarano, that from birth we carry expectation a priori with us as an element of our innate reflexes. Further on, we know that expectation a posteriori from the beginning is not preceded by expectation a priori. In addition, we know that expectation a posteriori could be connected to the newborn child's inherent ability to mimic the mothering persons'/another person's enactive motoric act of handling and kinaesthetic ability. An important issue worth putting forward is how the two kinds of expectation during ontogenesis were put together – their mutual relationships and their eventual "dynamics".

As should be clear from Chapter 3, my suggestion is that to be equipped with expectation a posteriori together with mimicking is a privilege reserved only for the human being. My suggestion also contains the idea that it is only the human being who has a wider exaptation to physical nature (far away from any other primate), with the extension of manufacturing tools with the creation of technologies in a seemingly never-ceasing flow. As should also be clear from Chapter 3, my suggestion also involves the assumption that without the equipment of expectation a posteriori, neither the emergence of *intention* nor the emergence of *imitation*, within the frame of operative development, would take place. With this capacity, the human has been furnished with an ability to learn, unprecedented by any other primate.

Back to expectation a priori and expectation a posteriori and their eventual mutual relationships/"dynamics". Thereby, my suggestion is that with the intimately connected emergence of intention and of imitation the two forms of expectation are amalgamated, during the subsequent ontogenetic development, in such a way that they both, together, work like one, in the service of operation. This does not exclude that expectation a priori keeps a certain autonomy in relation to expectation a posteriori. Let me illustrate this condition by an example.

We have the person who slips in the shower, loses his or her foothold and falls flat forward (I myself am such an example). In the fall the person finds the time to fend off bumps against the head by twisting the body lightly (the rooting reflex) and by doing so it is the right shoulder that hits the floor. I cannot understand this in any other way than that in the fall expectation a priori comes first, followed by the eventual surprise that lags behind.

It really has to matter – all different enactive motoric movements produced by the mother persons/others in dealing with all kinds of physical things/manufactured objects/artefacts – for the child from birth, and further during ontogenesis (see Chapter 3). We thus have to pay attention to, and distinguish from each other: on the one hand the physical things/ manufactured objects/artefacts; and on the other hand, all these different enactive motoric movements of the mother persons/others in dealing with the physical things/manufactured objects/artefacts. In addition, we have the mother persons'/others' enactive movements *without* any physical things/manufactured objects/artefacts. The important question is whether, and thus when, the child really does detect any difference between the *gestures* used *with* physical things/manufactured objects/artefacts and the gestures *without* such matters. Whatever the conditions, the child has in some way to receive, "interpret" the different gestures, and to put them in a context (and gradually "classify") – and this capacity is denied other primates. Michael Corballis and others with him would not be quite wrong in their interpretations of human gestures as one of the precursors of the emergence of language.

Back to the child and his/her world of a diversity of gestures. The exposure to all these gestures also implies a lively activity in the child's mirror neurons in the motoric cerebral cortex, AIP, and even in other parts of the brain (see Chapter 1). In the child's brain, there is an issue of passing through "constraints" in order to create interfaces and networks, following all passages through constraints, as a basis for continuously operative (and language?) functioning, at first on a sensorimotor level, and gradually on a cognitive level. The important issue of how to create interfaces will be my next task in the book.

The passage through constraints and also the creation and bridging of interfaces

The starting point is the proposal that *at the moment of passing through constraints, new interfaces are created, simultaneously*. To illustrate how an inquiry into the creation of interfaces might occur in analogue and digital systems, I start with a quotation from Anthony Wilden's work *System and Structure* (parts of the quotation are already cited in Chapter 3):

> A "bit" of information is used to select one difference from an infinite number of possible differences. [...] This difference then becomes DISTINCT from all the others. [...] Since the "bit" marks a boundary [...], it introduces some form of discreteness into the continuum [of the analogue]. (brackets mine)[7]

Furthermore:

> [Interfaces] to be crossed – it applies to the analog system too – always presume the digital. [Whereas], the analog can never be properly represented in the digital [except for the affect surprise-startle]. (brackets mine)[8]

John Haugeland has devoted special interest to the role of interfaces in systems thinking in general, and specifically when applied to the human as a psychobiological being. Interface is "a point of interactive 'contact' between components such that the relevant interactions are well-defined, reliable and relatively simple".[9] Moreover, he considers the interfaces to be of crucial importance to the human being with regard to: "intimate intermingling of mind, body and world [at the same time as distinguishing them]".[10]

The analogue can never be converted into the digital, on the basis of this postulate, I argue in my two latest books in Swedish that the digital machine, the computer, may never become human. The vital human

7 Anthony Wilden, 1972/1977, p. 159.
8 Anthony Wilden, 1872/1977, p. 461.
9 John Haugeland, 1998, p. 213.
10 John Haugeland, 1998, p. 224.

qualities are founded on the biopsychological basis of the onefold, which is more or less given. It can never be transformed into the twofold's either/or without harming the positive and negative affects, and the true creating and maintaining of internal presentations.

On the basis of the proposal that surprise-startle has elements of either/or, we may avoid the "digitalization of interfaces in the analogue" (for instance, with the creation of categories as a consequence) being attributed to a *deus ex machina.*

Thus, surprise-startle is the affect of transcendence at the same as it might be a contributor to the creation of interfaces. It triggers, normally, everything that we include in what we define as human functioning. This way, surprise/expectation (a priori and a posteriori) contributes at first to "the third essentially a reflex" (see the quotation above from Tomkins). This would cause expectation to be "infected by contagion" of the affect (surprise-startle) early on after birth.

Silvan Tomkins considers music to be closest to the affects, as they are shaped in human life. But in order to understand where both the similarity and the difference between them lies, we probably need to know more about the interfaces between them. In the connection with systems, interfaces are (as has been pointed out above, in Chapter 1) junctions – and, at the same time they are dividing lines between different systems, between system and subsystem and also between different subsystems. Interfaces are no "innovation" resulting from Alan Turing cracking the code for the universal Turing machine, converted into the modern computer.[11] They have existed in nature for billions of years, as features of the quantum leap; within the human being, I persist, in the operative functioning (through keeping the "event traces" in the abstract conceptual memory – "a spatial event that with a 'square' in space 'deceives' the space-time of its clock-hour").

Is the digital leap, seemingly with the thermodynamic features of the quantum leap, once and for all deemed to remain invisible and or incomprehensible for us? Within IT one speaks of "user interfaces" – which can refer to interfaces between different software modules and/or hardware

11 Alan Turing, 1936.

modules. Thus, the human being *knows* how to *create* and to make *use* of interfaces.

As an addition to Wilden's and Haugeland's definitions of interfaces above, I supply one from Andy Clark: "An interface [...] is indeed a counterpoint between two items across which the types of performance-relevant interaction are reliable and well defined [...] (and plenty of them [interfaces] within the brain)."[12] "Counterpoint" is used within music as the "tool" to interlace individual stems, and also each part with its melody, in a "polyphony". Johann Sebastian Bach was not the first, but probably the greatest master of composing this kind of "structural music".

If Wilden underlines the "invisible" (in the sense of difference being "no-man's-land") in the digital leap, with the formation of interfaces as a consequence, then Clark and modern computer technology emphasize the "usefulness of interfaces as bridges".

Becoming a human being: the necessity of the two affects interest and shame transforming into "one"

According to Silvan Tomkins, interest and shame are the two affects that have been overlooked (Chapter 3). During the first eight months of the child, interest/shame serves the child as a guide in his or her relations in the mothering world – with different grades of enjoyment towards joy as discharges. As extremes, they may take the form of: at one end an abundance of (and over a long period of time) high-grade discharging enjoyment with interest as its starting point (widened gaze); at the other end an abundance, relatively, of low-grade discharging enjoyment with shame as its starting point (lowered gaze). As pattern formations both extremes ought to be involved in the founding of the two fundamentally different prerequisites for "the formation of me-ness". If interest in its foundation is directed outwards towards the caregiver/other, then shame is without doubt directed inwards towards the child's entrance into "me-ness", in the

12 Andy Clark, 2008/2011, p. 33.

long run towards the child's development into adulthood. Shame is a clue for the individual inwards, in relationship to the individual outwards, outermost the individual as a social being. Present at each transgression – in social contexts as well as in innovative transcendences, or transgressions – is shame, as the threshold to be crossed. In this there ought to be an extension towards a specific *interest/shame script* in *Srip* as a basis for how we go on to relate to "the familiar" versus the "the strange". It might be described with an indication of direction in the following way: from within → out/from the outside → in; it is hardly wrong to describe it partly with what Nicholas Humphrey defines as "circle of causality". The grading and the quality in this motive formation in *Srip* probably is unique for every individual. The question one might ask oneself is whether this script in *Srip* is a fundamentally *directed* motive formation. My suggestion is that *the interest/shame script is such a primary directed motive in Srip*. May one at the same time regard it as a guidance for directed motive formations during the continuation of ontogenesis/epigenesis? As I understand it, the question about "directed motives" is a so far unasked question within psychology as an academic discipline. In the psychological literature there is a large number of definitions for pure descriptive usage: wish, longing, languishing, hanker, urge, desire, inclination among others. In my two latest books in Swedish I put forward suggestions about at least two directed motives that would emerge as superstructures upon the interest-shame script: *Wunsch*/desire and *Sehnsucht*/yearning. To that I will return.

The contagion of the affect as a supporter of imitation: the one-year-old

We are now approaching the decisive step towards the emergence of the child's *evocation* of the *absent* person (the caregiver/the stranger). In the ontogenesis of the child, the evoked word (the left hemisphere) has an advantage in relation to the evoked visual internal presentation (the right hemisphere). Could this be due to the advent of the capacity for *imitation*, and is it supported by elements in the *Srip* formation? In that case, we may speak about it in terms of something on the developmental line of affect/internal presentation underpinning an element on the developmental line of operation.

What we know about the well-oiled imitation and its importance for learning is that it is a direct emulation of a motoric pattern of gestures, or of a pantomime (that no chimpanzee in the world has the gift of managing) – and at the same time as a feature of the motoric complex uttering of words? Earlier I have described the infant's deeply embedded inclination to put all moderately big things (hard as well as soft ones) in the mouth in order to suck/chew them, and that this probably has to be "a sort of mouth gymnastics as a precursor of word uttering". Here I would like to insert an observation of Jonathan, thirteen months that afterwards aroused a strong wonder in me. He was just learning to walk (although still a bit unsteadily). During the three hours I observed him, he did not put a single thing (neither soft, nor hard) in his mouth. He picked up a new little present, a crocheted and upholstered cuddly toy, and held it against his cheek. Later on, I had it confirmed that it had become a rule (not without exceptions) for Jonathan not to put things in his mouth. Time to utter the word for the thing (except "mam", "dad", "thank you"), instead of sucking on the thing in the fivefold raising?

As a basis for the capacity for emergence of *visual evocation*, I suggest as necessary elements: the contagion of affect, the emergence of recognitive internal presentations, imitation and the evocation of the first word. Teresa Bejarano equates the ability of evocation with "symbolic ability" (Peirce), in which she includes the pointing gesture. From, in my parlance, *mimicking*, the one-year-old now passes on to *imitation* ("simulation", Bejarano) of a "model" – an "adaptive benefit" that "sets the child free" from only perceiving "a motor sequence of movements, and only to imitate kinesthetically". "Imitation (simulation) of a model's motor sequences is a precursor for the imitation of articulatory-phonetic patterns … ."[13] Bejarano pays great attention to the transition from "mimicking of motoric patterns" to "imitation of a model". What, then, according to Bejarano, does "model" stand for? In my account I find the following descriptions: "agent in an earlier scene"; "agent that is not myself"; "agent as distinguished from the real world". These are unusually abstract descriptions by the author, for

13 Teresa Bejarano, 2011, pp. 119–121.

whom details are of great importance. Mind you, Bejarano's project is to derive the syntax of language from the child's "innate cognitive equipment".

Bejarano and I are concordant in the opinion that mammals may have access to recognitive internal presentations – but hardly to the ability of evocative internal presentations (not even the chimpanzee). Bejarano's derivation takes place via sensorimotor ability, imitation of models, the articulation of the word ("the linguistic symbol") and the make-believe ("symbolic play", see Piaget). To my mind, Bejarano disregards in this case the importance of the attachment/relation and the affective communication in connection with the caregivers.

After the digression from Bejarano's derivation, let us go back to the development of *Srip* and its possible meaning for the emergence of the ability to imitate a – "model". "The model" that I refer to is the "familiar versus the strange 'agent'". At the time when this polarization has taken hold of *Srip*, recognition has opened up to embrace the understanding of the "constant" caregiver as well as of the "other" to be beings having the dynamic inherent polarization of "the familiar" versus "the strange", and what the continuum between the polarizations includes. To the one-year-old, the polarization is determined by the contagious experience of the affective communication in the mothering world (between the poles: abundance of positive affects at the expense of negative affects – abundance of negative affects at the expense of positive affects). This ought to be decisive for the one-year-old's tuning into the caregivers as well as to "others".

A cardinal feature of the affects is that they are contagious, which probably contributes to both immediate displacement and condensation in *Srip*. But displacement and condensation, presumably, first and foremost have to do with the affects' overarching characteristics of being *general* and *independent*. This implies that the affect may "be invested in other affects, combine with other affects, intensify or modulate them, and suppress or reduce them".[14] My suggestion concerning the mutual influence affects have on each other is that this influence first manifests itself in the synchronously functioning *Srip*. This is worth a more elaborate and richer account of what I shall return to below.

14 Silvan Tomkins, 1991, p. 72.

Here I return to the characteristics of contagious. My proposition is that *contagious functions as a fundament for imitation*, that appears at the age of one (see above). For with *Srip* the child has got access to a (primary) psychic system to handle affects as well as to exercise influence on the agency register that is triggered by expectation from within – but not to forget, what the ongoing course of events has to offer. In sum, the one-year-old has access to both expectation converted into *intention* and affect converted into *emotion* (the latter on its way towards the entrance as motive in the service–disservice of the subject – disservice against the backdrop of my experiences with clinical work), in the shape of emotional script patterns. My proposition includes the idea that the emergence of imitation is made possible by mimicking on the line of operation and by contagiousness on the line of affect/internal presentation, thus, as being substructures. How may one otherwise understand the smooth adherence of imitation in the perception and the imitation of, on the one hand, the complex pattern of gestures and pantomimes, and on the other hand, the emotive patterns that appear to the child on the continuum of "the familiar – the strange"? Obviously, with these features in their place, there is a particular preparation for the "linguistic symbols", as early converted into *evoked* and phonetically ordered expressions as they are in ontogenesis – but not yet converted into phonetic expression in inner silent speech.

Brief account of the obstinate age

From every parent of a two- to three-year-old child, you may hear that their child is in the obstinate age. What is it then that the child defies and what is defied? What we do know is that the child has attained sufficient control of language that it may be used in the service of obstruction. It is common that developmental psychologists describe that when the child acquires a new capacity it has a tendency to "overuse" this capacity – the capacity in the beginning being both "omnipotent" and "vulnerable". It is in this phase that there are both constraints to be crossed and interfaces to be developed. In each of these phases the child is particularly vulnerable to the "resistance" the child may distinguish in the environment/the caregivers.

The period of the obstinate age (two to three years old) is defined by the psychoanalyst and childhood researcher Margaret Mahler as *rapprochement* – the child is on its way to separating him- or herself from the caregivers "out of sight", but the child has not yet reached the capacity for "emotional object constancy". Therefore, the child constantly has to return to the caregivers in the room in order to confirm their "real" existence. Originally this is Mahler's parlance, but I have converted into mine, given the reference in the affects of interest/shame, which are in their way building a specific emotive script in *Srip*. In this state of ambivalence – Mahler's term is "ambitendency" – the child oscillates between infantile clinging and fierce repellence accompanied with a "no" – obstruction.[15]

The big leap: the age of three years

Now, how does Teresa Bejarano express herself when it comes to "desire", "objectives and instruments" and "adaptation"? To facilitate the reading, for you as a reader, I will repeat from Chapter 3 that the access to any form of desire is an adaptive availability only when there exists "an instrumental objective" tied to the desire. In my interpretation above, I understood Bejarano's turn-taking as that there was first an (evolutionary) emergency of "the instrumental objective" that eventually was in need of an access to the emergence of "desire as motive" ("the cognitivist mistake").

As we saw in Chapter 3, what takes place within the development of the three-year-old is complex, with the emergence of a paradox. I start with an account of what we know about that: operatively, imaginatively, affectively/emotionally, relationally and with regard to identity or "meness" formation.

Operatively the child starts to comprehend that in sets include separate elements – the balls in a drawstring pouch – possible to distribute "fairly" to other children. The three-year-old child gets access to addition according to the principle of $n + 1$. This the child may convert into things in the

15 Margaret Mahler, 1965.

room with the aid of the numbers 1, 2, 3 – that successively embrace larger and larger amounts. At this point the three-year-old already has access to subtraction according to the principle of $n - 1$: If one subtracts a spoiled and discarded apple from a pile of apples, the child knows that there is one apple less, and that it is forever gone. Subtraction precedes addition in ontogenesis. What we also know is that the three-year-old starts to distinguish between time and space without the time order or the time range forwards or backwards being established.[16] To my mind, these achievements require access to the primary operative cognitive ability. The latter I will illustrate by giving an example that clearly manifests itself in Jonathan, now two years and eleven months old.

Jonathan and his family come to visit grandmother and grandfather, where they live during the summer. Grandfather gets the watering can in order to water the flowers in the pots and baskets. Jonathan asks his grandfather for the watering can, because he wants to water himself. Everything goes very well. At the last pot with the blue-and-white daisy bush he waters the flower – but directly after his grandfather (the author) saying "now it is enough, I think", Jonathan throws the can at the plants. Thereafter he looks at his grandfather's face and eyes, and fixates his gaze there for an instant/Augenblick. The grandfather reads off a glimpse of "wonder/appeal". The grandfather comments: "The flowers probably do not like it."

The question is: What lies behind Jonathan's throwing of the can at the flowers? Is what caused it "affectively" or "cognitively" conditioned? After examining the questions thoroughly that come to my mind, I would answer them as follows: The vantage point is the idea that Jonathan just entered the introductory phase of acquiring "cognitive operation with reversibility". But he still lacks the ability to retain the reversible element on the level of thought, with the consequence that the reversibility is implemented on the (sensorimotor) "level of agency". With grandfather's gaze of "wonder/appeal", Jonathan searches for what is to him the obscure affective/emotional/feeling/moral implication of his "counteract" in the gaze of his grandfather, in *retrospect* (*nachträglich*). Conclusion: Jonathan's

16 Teresa Bejarano, 2011; Alison Gopnik, 2009/2010.

double "act" – "watering the flowers is 'a good thing' as an 'operative act of thought'" and "throwing the can at the flowers is 'a bad thing' as an 'operative act of thought'", is in fact founded in the primary operative cognition with its embryonic reversibility, which in this early phase expresses itself as a "counteract". And in this phase of ontogenesis the "delay" (see below) of the "inner life" – affects/emotions/feelings/moral – becomes obvious in relation to the understanding of the "outer life" – the consequences of the operative "act of thought" in relation to the outer world.

What is crucially new on the *line of internal presentation* is the emergence of *visual evocation* of the absent object. Nobody knows for sure – but I'm inclined to share Bejarano's opinion that such an evocation is reserved for the human being only – "A big step for sapiens".[17] If we convert the visual evocative internal presentation into the notion of "rheme" in the development – does it have a necessary "predicative theme"? In my answer to that I refer without doubt to *Srip* as "theme". Is there any "predicative" element in *Srip* that points to "rheme" in the form of evocative visual internal presentation? If we follow Mahler, the necessary precondition for the emergence of "emotional object constancy" is that the child can evoke an internal presentation of the "familiar other", who, according to the child, may express *both* positive ("good") *and* negative ("bad") emotions; but eventually also of the "strange other".

Lev Vygotsky has a general insightful formulation of the importance of affects/emotions/feelings for the evocation of internal presentations, "the phantasy life":

> Psychology calls this influence of the *emotions* on combinatory fantasy the law of the general *emotional* sign. The essence of this law is that impressions or images that have a common *emotional* sign, that is, produce *similar emotional* effects in us, have a tendency to cluster together, despite the fact that there is no association among them either based on external *similarity* or *contingency*. A combined product of the imagination is generated, which is based on the common *feeling* or common *emotional* sign uniting these diverse elements that have become associated [...] *Affective similarity* unites and ties together objectively *dissimilar* images. (italics mine)[18]

17 Alf Nilsson, 2005, p. 172.
18 Lev Vygotsky, 1933/2004, pp. 18–19.

I see the possibility to convert this formulation in a compressed form into the dynamics of the emotional scripts in *Srip* (see above). As I understand Vygotsky, what he sometimes defines as "emotion", sometimes "feeling", sometimes "affect", is not anything "reactive" but that which triggers the evocative internal presentations/"phantasy life" – "phantasy life" is driven by "motives". How such a motive may appear, I shall soon return to.

The common formula for affects and the relations is converted by the contingency principle S-R-SE (Someone in Relationship to Someone Else). The great scientific contribution of Sigmund Freud was the discovery of the psychic system *Ucs* with its "fivefold logic". The fundament of it was the derivation of the relational development of the child, in what he denominated the "Oedipus complex" at the age of three years. Nowadays it is translated into our common experience of the three-year-old that in a transition period perceives the caregivers as "mother" and "father" as more "dissimilar" than earlier; and that for the girl the relation to the "father" is more "interesting" than the relation to the "mother", and vice versa for the boy. It hangs together with the fact that the child's "me-ness" or "gender identity starts to stabilize" (but not until the school age, the child will understand the role of the genitals in distinguishing the genders).[19] Moreover, at this point the child starts to shift from referring to him- or herself with the name to the usage of the pronoun "I". For a comprehensive comment on the development of the identity I refer to P. Mitchell and F. Ziegler.[20]

Desire as directed motive

I will initiate the discussion with a quotation from Anthony Wilden, who I think leads us in the right direction towards what *desire* connotes in the human world – the only world in which desire is to be found (desire is not to be found in the chimpanzee world):

19 Susan Coates, 2006/2009.
20 P. Mitchell and F. Ziegler, 2007.

> Desire comes into the world because of the necessary relation with others that makes humans human; the child's helplessness to attend to his own needs results in the detour of need, [...] and it is this detour which generates *unconscious* desire as a fundamental and unfillable "lack". (italics mine)[21]

Is there any trace of the affect surprise-startle in *Srip*? As a neutral affect and the whole organism's readjustment affect – not least, being an important affect that triggers expectation (see above) – it ought to activate both the analogous "continuums" and the digital "bits". Moreover, I've suggested that surprise is involved in the amplification of stimulus – but not in the response, the latter's value always being marked by the positive and/or negative affect. Early on in the book I describe *Srip* as dynamically structured by affect derivations and sensory experiences. I have ascribed the affect script interest/shame a particular importance as a directional script in *Srip*. More generally, this concerns a stimulus amplifying positive or negative affect that via amplification of the response gives a marking of positive or negative affect value, which fundamentally is the affect script interest/shame converted into an emotional script. This should turn *Srip* into a psychic system with a strong affect potential for recognition of the "familiar" versus the "strange"; in the phylogenetic perspective, the eight most important domains that once were crucial for the survival of *Homo*. For *Homo psychicus* of today, it is always a question of a re-presenting (recognition of the interest/shame script). *Srip*, thus, is not an ignorant and absolutely not an agnostic, but is characterized by a dynamic preparation to mark along a continuum, more or less tumultuously, with similarity at one end, all the way out to the edge at the other end of strangeness/dissimilarity.

At this point, we know that affects are, if we want to define them as such, "the letter elements" in the infant's communication in/with the mothering world. The nourishing, the sometimes thinned and the sometimes overloaded consuming affects as elements in the communication appear through the *relationships* (to the caregivers) as "supporters". In the re-presentings of the affects, converted into emotional scripts, the relations, thus, according to the contingency principle S-R-SE, are crucial for how

21 Anthony Wilden, 1972/1977, p. 23.

the scripts are woven together. Usually, in the world of humans, relations are unfolded, as mentioned, in the force field between the pole of interest and the other pole of shame – interest and shame are the *relationship affects* par excellence. Now we know that shame is triggered by incomplete interruptions of interest – having other negative affects on tap if interest, in perpetual readiness to close the break in the relation, should succumb. It might be of some value to repeat: this makes interest and shame a dynamic complex in the human world, beyond the prerequisites in other animals' ecological niches. Shame, in terms of "by humans, for humans", was something that Darwin was aware of when he wrote:

> Blushing [under a sense of shame] is the most peculiar and the most human of all expressions. Monkeys redden from passion, but it would require an overwhelming amount of evidence to make us believe that any animal could blush.[22]

Before continuing I would like to take the opportunity to interpose a reminder of the fact that the caregiver who, to the child, might be described as a "dyadic hybrid": sometimes *equipped* with physical things/manufactured objects/artefacts, sometimes *without* such equipment. During early childhood, the child has access to a transitional space with room for certain manufactured objects as "comforting objects with double reference" – one of the "ends" with reference to the mothering persons, and the other to the manufactured objects, the teddy bear, the comfort blanket etc. – which Donald Winnicott called "transitional objects".[23] One function that possibly could be ascribed to the transitional object is the bridging between the analogue and the digital world, fundamentally between surprise and expectation – but also between the affects of interest and shame. Such a bridging could also imply, through surprise/expectation, that anticipatory elements of persistent and rampant extension in space are added to *Srip* – when, in the development, the time is right for keeping the synchronous emotions (surprise/interest/shame) "event tracks in mind". For

22 Charles Darwin, 1872/2009, p. 286.
23 Donald Winnicott, 1953.

the three-year-old child still lacks access to time as evocative extending backwards and forwards.

Henceforth I'm willing to interpret Teresa Bejarano's linkage between "desire" and "the instrumental objective" with benevolence. I primarily focus on "desire". I take as my starting point the interest/the shame complex's conversion into *Srip* as guidance – the key to the child's further development into a relational, societal and social being; inserted in the evo-devo perspective, to the child's development into a relational and teleodynamic being (Terrence Deacon). We have seen that the child at the age of three years is subject to what is, in many respects, a substantial leap in the development. It is also at this age that the dominance of the right hemisphere ("the hemisphere of the affects/imaginations/visuals") develops and crosses the dividing line towards a transition to a dominance of the left hemisphere. This is illustrated with a figure in some of the earlier books – worth mentioning in this connection (Figure 4.1).

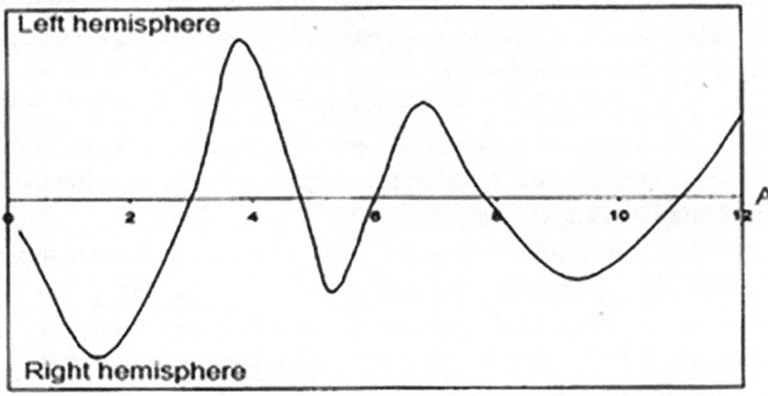

Figure 4.1. Both hemispheres' dominance throughout the different phases of the development of the child, during the first twelve years of infancy. We see the intersection at the age of three years towards the first dominance of the left hemisphere (the hemisphere of cognition) in relation to the right hemisphere (the hemisphere of affects). The figure is extrapolated from Allan Schore's illustration, 2003.[24] The figure is also to be found in Alf Nilsson, 2009 and 2013.

24 Alan Schore, 2003.

If Bejarano's thought is that to "the instrumental objective" the ability for evocation is added – not unreasonable given that she regards the ability to evoke (together with imitation/simulation) to be a driving lane to "the other mental centre", through which the evolution of the human line was outlined – the thought that some type of re-structuring of the organization of motives takes place, converted into a new ordering of motives, does not seem unreasonable. Interpreted in this way, the emergence of "desire" is a consequence of such a re-structuring/re-presenting.

For the three-year-old, at least, a clear transition occurs from a fundamental system formation in the shape of *Srip*, to a seemingly new formation on the primate line in the shape of something that is reserved only for the human being, namely the System of Evocative Internal Presentation, *Seip* – an extension implying that the outer absent object may be evoked as a *visual internal presentation* within the child. My suggestion here is *that each new ability on top of Srip, staged as a new system formation on the line of affect/internal presentation during ontogenesis, has to be motivated* (not only by me formulating the proposal) *with a motivation source newly formed for the human being* – according to the principle of evolution that says "you take what you have": like snakes that million years back kept their jawbone to the ground to register vibrations; so did evolution by transforming the jawbones of the snakes into ear bones in the ears of the mammals.

What I have so far underlined is the interest/shame complex being an emotive script formation in *Srip*, and with a tendency to function in a targeted way. Fundamentally, this is about a positive gradient affect directed *outwards* to the world, along with an auxiliary affect (auxiliary on top of the original affects fear, anger, distress and, in addition, dissmell (see Nilsson, 2005 and above), and also a relational affect directed *inwards*, shame, towards "the me-formation". What we also know is that the child's history of past relationships, containing a mix of interest/enjoyment as nourishing elements, and shame/enjoyment as depleting elements, is a factor to be considered with regard to the three-year-old's ontogenetic leap. I want to remind you here that interest/enjoyment ought to contribute an abundance of a discharging high-grade enjoyment-joy; shame/enjoyment, on the other hand, with an abundance of a discharging low-grade enjoyment. There is no three-year-old that has not experienced both but, as already mentioned, in different proportions. Thus, I've reached the generic proposal that: *each new order of motivation on the human line is an order of relational directednesses – outwards* (interest) *as well as inwards* (shame), *both anchored in the "well-known" and in the "strange" (with reference in Srip) – that are always present as an underlying compass for the sake of orientation.*

If we, for a moment, restrict the discussion to focus only on the relations of the three-year-old (to the caregivers), it was Sigmund Freud that put us on the right track, in my circumscription: for the girl, an augmented interest in the relation to the father and augmented shame in the relation to the mother (identified as being "of the same sex and as such" as someone being as "me"), and vice versa for the boy. How, then, are we to understand the positive affect of interest directed outwards, and the negative affect of shame that is directed inwards? This may only be comprehended in terms of being new formations in the human world. As is illustrated in Figure 3.1, I consider the human as being a drive/need being, described in detail in my first book in Swedish about *Homo psychicus*. If we disregard here the necessary drives/needs (need for breathing/oxygen, thirst/liquid, hunger/food) and solely concentrate on the sex drive, it has traits, from early on in the development, of "megalomania/omnipotence/narcissism/monomania" – where shame is directed inwards, with its negative feature, and strives to keep this feature under the control of shame and admonition (but with the gate half-open to the positive affect of interest).

Specification of desire as a directed motive

Now it may be time to for a more precise proposal about one of the new motivation orders that are solely reserved for the human being – and this time situated somewhere *in the interface between Srip and Seip*. As a reference to this proposal I adduce (i) Anthony Wilden's thesis about *desire* always being directed towards relationships; (ii) Teresa Bejarano's opinion that *desire* is always adaptive if it is tied to an "instrumental objective" – the instrument that, apart from pointing, she allocates to, *inter alia*, the emergence of *visual evocation of the absent object*; (iii) Sigmund Freud who early on described *Wunsch* as being linked to the emergence of the "Oedipus complex" – Wunsch, in the *Standard Edition* first translated with *wish*, and then with *desire* – as "nothing else [...] can drive our *psychic* apparatus to work" (italics mine). When Freud later on formulated his topographic model (*Ucs*, *Pcs*, *Cs*), he allocated desire to the psychic system Ucs or to the "border between *Ucs* and *Pcs*". Later on, Max Schur elaborated Freud's derivation of desire as always being directed to "a specific object" ("object" in the psychoanalytic sense) defined as "the Oedipal object". A more detailed account of this is to be found in my early book on "defence mechanisms/wardings off". For a discussion on how I argue for the many overlaps I find between *Ucs* and *Srip*, I refer to my second book in Swedish about *Homo psychicus*.[25]

Desire as a directed motivation order exists in the human being – but, I repeat, not in any other living organism on earth. What we know so far about the motivation order is that: (i) it is bidirectional, pointing at the same time outwards (based on the affect interest) and inwards (based on the affect shame); (ii) primarily it is directed towards relations; (iii) as an interface it contains digital features, what we may argue ultimately, with traits of expectation based in surprise and thus, if not with an endless, then at least with an indeterminate extension towards the outer world; (iv) simultaneously with having an extension towards the inner world via the emotive shame/enjoyment script it ought, in certain persons, to be close to actualization of other negative emotive script formations. To the latter I will return in Chapter 6, in which certain psychopathological states that can appear in *Homo psychicus* will be described and illustrated with the aid of PORT.

Here, however, a small digression is in order to anticipate what is to be dealt with later on in the account. To desire at least one more directed

25 Sigmund Freud, 1900/1996, p. 510, 1915b/2003; Max Schur, 1955; Alf Nilsson, 1983/1994, 2009, chapter 2.

motivation order ought to be added as a bridge (preliminarily derived and designated in the two latest books in Swedish). As a bridge, it probably has its seat in: most deeply surprise, coloured by expectation; the interest/ shame script with a tendency towards directedness; the directed motivation order desire.

Towards Sehnsucht/yearning in the light of the evolutionary perspective

For some reason, the human seems – as even science does, for that matter – to be perplex when it comes to differentiating and describing the most advanced motivation orders of the human being. As already mentioned, we have a variety of descriptive denominations used in different ways by different authors. May it be connected to the fact that it is interest (fundamentally) that is directed to *the outer world*, and the content there, that primary preoccupies the human being? Furthermore, that it is fundamentally shame, as a "warning signal" inwards that is kept aside or, is made an exception? In the wide perspective of evolution, it was adaptation to the outer world that once preoccupied *Homo*, something that continuously dominates *Homo psychicus*. Earlier I've described it in terms of there being a systematic delay in the human's understanding and dealing with his or her inner world – at the most in some shape evoked in a form of "me-ness" – in relation to the outer world, where the operative cognition has directed its laser beams: all the way out to the outer macrocosm, as well as all the way in towards microcosm.[26] I regard it as the human tragedy that man is so "retarded" when it comes to the comprehension of the inner cosmos at the expense of the outer cosmos – the inner cosmos expressed in the shape of psychic/mental "language" (the knowledge about the neurobiological processes behind can never be translated to this parlance). As a reader, you may perhaps object: "Literature and art are much better at giving this a language!" Indeed, that is a type of "language" that I would not like to be

26 Alf Nilsson, 1983/1994.

without. But are we not in a phase where we need both, thus, to understand the inner life through science as well as literature?

To this it may be added that the human being, in the operative/mental and linguistic aspects, is almost unbelievably advanced considering the fact that s/he is fundamentally an earthly being. Seemingly, the human being is indeterminable when it comes to where the limits for our knowledge about the macrocosm as well as the microcosm are (we are still left in indetermination on the level of quantum mechanics/dynamics).

According to the legend, the command above the door to the temple of Apollo said: "Know thyself!" How far have we come, in fact, in the process of getting to know ourselves as persons, relational, and societal beings, and also world citizens? We all know that the precondition for life on earth – which began more than three billion years ago – is a thin biospheric film that surrounds the earth. We also know that we human beings senselessly bombard this film with greenhouse gases that consume the biosphere and the earthly climate with unforeseen consequences. Unforeseeable already within a century. In 3,000 years, these consequences will be even more unpredictable, that is, from our standpoint, the Western civilization process here considered in the same timespan but in a future perspective instead of a historic one. It is as if we still were living in the Platonic/deistic perception of time as being endless. Before Plato, Protagoras (481–420 BCE) formulated conditions on earth, using Man as the measure of everything, and thus, he became responsible for life. Why do we still know so pitifully little about an exclusively inner psychic phenomenon as *responsibility* – and the extension of responsibility? Responsibility – only reserved for the human being on earth to experience. My proposal above is that responsibility belongs somewhere on a dimensional scale which starts with *shame as affect* at one end and ends with *guilt as feeling* at the other end. If this is so, "responsibility" is squeezed in as a paltry element on a negative affect dimensional scale, starting in biology and then ending as converted into perception within the human being. In a world guided by words like economism/profit/individualism, responsibility – extended to a dimension of collective responsibility and solidarity, and also to responsibility for the survival of life on earth – seems to be washed away.

After these digressions – back to the three-year-old having a desire, in the form of an upgraded superstructure and extended motivational order. Considered in the light of relations, desire probably is the precondition for both infatuation and love; love that preconditions at least one more superimposed motivational order (see below) – love, belonging to the human being only. Desire contains features of rampantness, having expectation as a foundation; via the affect of interest as tonic (see Silvan Tomkins above), which triggers the augmentation of the sex drive towards excitement; with influence from the affect of shame, a door opens half-way to deliberation; with the features of both interest and shame, openings to *illudens*, delusion, are triggered, as well as to *ludens*, playing. As we have already seen, desire transformed outwards into relationships, and inwards into "me-ness", opens a transition towards gender as well as towards dissimilarity; towards personhood as well as towards the dissimilarity "I"–"you". Successively, during development, it opens towards everything and everyone. As already mentioned, the emergence of at least one additional motivational order during ontogenesis is added to this. Jean Piaget does not give, to my knowledge, the specific operative development that takes place at this age (around three years) a precise name. By starting to handle addition and cardinal numbers consecutively towards "infinity", a big step is taken (the development of concrete operations takes until the school age before it starts and the child may engage in multiplication and division). In my opinion, the important operative addition to the three-year-old consists in the ability to perceive space as one thing and time as another. Above I speak of the addition to the operative line in terms of "primary cognitive operations".

Outmatching everything else in the three-year-old is the ability for visual evocative internal presentations. Considered in the evolutionary perspective – and in connection with that perspective I agree with Bejarano – it was an absolute crucial addition to the primate on its way towards becoming human. The distinct and cold operating with the world, as if it was a thing, has led us all the way to the edge of the abyss where life on earth and the biosphere are today. To be able to be close to "the Other" within oneself, and to "the Other" elsewhere, has driven a wedge of responsibility between the outward heat of interest-excitement, and the inward self-destructive welding flame of shame-humiliation, and the feeling of guilt as a painful state, but in spite of all, with "contemplation" as an inner psychic possibility. It is worth repeating that desire is a result of a digital interface formation. It has elements of both opening positive and closing negative feedback. To this we may add that the interface of desire is transparently bottom-up – in the adult bottom-up as well as top-down (Wilden). Finally, desire always seeks the "absolute satisfaction", but still has to settle on what the world and life have to offer. Desire alone has no qualification to trigger "transcendence".

Affect and internal presentation in the shadow of the operative cognition in the child's continuous ontogenesis ...

If we let ourselves be guided by Figure 4.1, what follows now in the development of the child is a period dominated by the left "cognitive/linguistic hemisphere". How is one to understand that? Michael Gazzaniga (specialized in the area of *split-brain*) gives us a certain guidance to the answer to that question: "The right hemisphere maintains an accurate record of events ('the affect always speaks the truth!'), leaving the left hemisphere free to elaborate and make inferences about the material presented" (parentheses mine).[27] One way of interpreting Gazzaniga's neurobiological description, transformed into the development of the child after three years of age, is that the "cognitive operative/linguistic domain" is in a phase of integration of the multifaceted developmental leaps that have preceded – for "updating" and "decision control". As I have demonstrated in detail in my fourth book in Swedish, the child at this age and, continuously until the school age, is still deeply drawn into the pretence play. Changes take place in the imaginative, described in this book – but the child still lives with a disorienting sense of time with regard to *when-in-time backwards* situations and events occurred, and to the implication of *when-in-time forwards* (with death as a final point on one's own timeline). It is space that sets the time limits of the child. The perception of time is still set aside, compared to the perception of (earthly) space (Alison Gopnik).

27 Michael Gazzaniga, 2009/2009, p. 296.

As appears above, we know, through research by Jean Piaget and the School of Geneva, a great deal about the dramatic development of the child from the age of six or seven up to the age of eleven to twelve. In spite of the "neo-Piagetean critique" of Piaget's and the School of Geneva's "thinking in stages" one cannot deny the worth of thinking in terms of "developmental leaps". It is in line with modern evo-devo thinking and Terrence Deacon's "incomplete nature" converted into terms of thermo/morpho/teleodynamics (Chapter 2). One cannot escape either that: in research, in the wake of evo-devo thinking, cognitive and linguistic development have been in the foreground in a one-sided way. Possibly it is connected to our limited phenomenological ability as adults to pin the escaping affects/emotions/feelings down in operative sustainable *classes*, transformed into *notions* according to the ordering of *binary differences* ("brute facts"). Science is not always governed by curiosity, backed up by interest and desire (and also Sehnsucht/yearning, see below). It is also governed by domains where curiosity may be converted into irrefutable "scientific methods", today transformed into narrow-gauge and simple evidence-based thinking.

Sehnsucht/yearning inwards as directed outwards: our most advanced motivational order?

Sehnsucht in German/Yearning in English/*Längtan* in Swedish. In a definition of *Sehnsucht* in German one finds, among other things: "ein inniges [schmerzhaften] Verlangen nach einer Person [...], einer Zeitspanne [...], Todenssehnsucht". A definition of *yearning* in English is: "deep longing, especially when accompanied by tenderness or sadness".

It is not until the child reaches school age that he or she reacts with, and afterwards consciously registers "surprise" within him- or herself. This is illustrated in a study that Alison Gopnik reports. Children of different ages underwent an experiment. The children were sitting in a room without anything happening, apart from a bell ringing every thirty seconds. It happened repeatedly that the bell did not ring without anyone commenting on that. After a while, when the children discovered that the bell did not ring, they reacted with a delayed expression of the affect surprise-startle – Gopnik's term is "*startle*". When the children showed this expression, the play leader asked them what they felt and thought when the bell stopped ringing. The preschoolers answered: "Nothing!" "Were you thinking on the bell ring?" "No, [...] Every time you think for a little while, something goes on and something goes off. Sometimes something goes on for a couple of minutes and then there is nothing going on."[28] Children of school age and older children answered that they were "surprised". Gopnik interprets the results in the study as showing that the preschoolers cannot place a recent event, "the surprise" – as an expression within themselves as a consequence of the bell ringing having stopped – right in time. But that can hardly be the whole truth. The children did react with an expression of surprise-startle! So, in some way there was within them a sort of "knowing" of time (between two rings of the bell) that after a while caused the expression for surprise-startle ("startle"). Gopnik's interpretation is biased in favour of the cognitive: *cognition* within the child has not yet had the time to develop sufficiently when it comes to orienting the child concerning the timely dimension of an event backwards in time, or in the correct chronological order.

With an anchoring in the two lines of development that are considered in this book, a number of question are brought to the fore which I think are worth asking. The first question remains within the narrow operative cognitive domain: May it be that the affect domain, via surprise-startle, was trigged within the preschooler, but without expectation (the operational domain) – so far – being equipped with a real cognitively operating ability (concrete cognitive operations) to be triggered? Now I will move over to the domain of affect visualization to ask the next question: May it be so that the lack is here – because surprise-startle only has evocative internal presentations to trigger, but these are not yet transformed into evocative internal images and/or feelings to trigger? According to an earlier definition in this book it is the immediate evocative internal presentation linked to the immediate recognitive internal presentation, both in one, that are the foundation of *perception* – according to my theory of perception as

28 Alison Gopnik, 2009/2010, p. 151.

being two-layered, "irreversible" and that are fully structured on the visual plane from around the age of three years. The diachronically time-bounded perception lacks ordinary access to "retrospection" (for exceptions, see Chapter 6) and can never contribute to the act of giving "surprise backwards in time" (as in the experiment) a place in the preschooler's memory (neither in adults if it does not take the shape of a feeling?).[29]

Perhaps the reader wonders about the role of language in this connection. The preschooler does have access, at this point, to a rather advanced language: Could not language involve itself and "fill in the time gap" between surprise, as an expression, and the point in time when the play leader asked about what the child felt/thought? My answer to that question is: Everything indicates that language, within the preschooler, lacks sufficient "semantic content" to give an implication of time between surprise-startle, as expressed within the child, and the point in time when the play leader asked his or her question. Possibly included in this condition might be that the preschooler lacks access to *inner silent speech* (Bejarano, Gopnik).

Already here I put forward my proposal that what the preschooler most of all lacks, and which contributes to the failure of the task in the study, is a necessary inner motivational order. But to start from the beginning, I will return to "the particularity of surprise-startle" and surprise-startle as "the whole organism's alarm bell".

Earlier I've used an example taken from Nicholas Humphrey.[30] He lets the amoeba represent early evolution, and Man late evolution. Stimulation on the surface causes the amoeba to react without the stimulation taking the detour via any central nervous system. The reaction, and "sensation", the latter my contribution, of the amoeba is almost simultaneous with the point in time for the stimulation. Stimulation of the surface, of the sense organs, of the human being, always takes time – the activation of the human autonomous and central nervous systems always takes some time – and time thus becomes a feature of the human "sensation". According to the description above, surprise-startle is – when stimulated from within as well as from the outside – the inborn feature that triggers: on the one hand, one

29 Alf Nilsson, 2013, chapter 5.
30 Nicholas Humphrey, 1992/1999.

or several of the positive and/or negative "value affects"; and on the other hand, the "value-neutral" expectation, a priori and/or a posteriori. Each human being with experience – which, as we know today, also includes the newborn child – has to go along the stimulation all the way through the nervous system, where it is processed, before the response comes. Nowadays we know also that this process is connected to both positive opening feedback and negative closing feedback, the latter defined by Terrence Deacon as "constraints". The question one may ask is whether surprise-startle in some way influences, partly, the *when-in-time backwards* and *when-in-time onwards* (see above) during rest of our lives. In that case it would only concern the line of affect/internal presentation. Because when surprise-startle finally triggers expectation, the digital operative steering takes over completely and leaves surprise-startle to its own destiny. My suggestion is that on its way to the steering of the *when*-direction backwards and the *when*-direction onwards, it is the two gradient affects, fear and interest, that have the strongest influence. Thereby, regulation according to the analogous principle prevails, with an extension in time from synchronous to diachronous, at its extreme, the "timely state" of perception and emotion.

Let us continue the discussion by following, first of all, the way of expectation that in a timely sense is given from the beginning: steered by the digital principle, the operation follows a more or less traced path, with its specific developmental steps forward to the formal cognitive operation, or the reflecting abstraction. All the way along, the negative feedback has to ensure that the redundancy is kept under control, so that operation is not infiltrated by noise (is it such an infiltration which has arisen in the schizophrenic psychosis with its "distortive thoughts"?). One of the shortcomings of the thought/operative cognition ("the lack of thought", see Nilsson, 2011) is thus that it does not contribute to our subjective time perception, nor to how we handle our relation to the personal death. As appears in my third book in Swedish on *Homo psychicus*, it ought to be of a concern to the affect-imaginative course of development.[31] This claim, I'm nowadays inclined to clarify. With access to the language of mathematics,

31 Alf Nilsson, 2011.

we may deal with time as being measurable and at the same time being an "abstract" factor to count on. To measure time in this way is only possible for operative cognition.

But I'm staying put on earth, and sticking with the idea of time as a unit of content for the human, as an "earthly being". In my fourth book in Swedish I described this with a metaphor, "time as the invisible reverse side of space" (see also above). Each time we try to fixate our eyes on time (I disregard the clock as an instrument of measure, with time units created by Man as a cognitive being), the gaze ends up in space. Time in the perception as an ongoing evocation of now/Augenblick follows: firstly, the synchronous, eventually the diachronous path – that never reaches the asynchronous/asymmetric/digital position where time and space can intersect (Albert Einstein).

The argumentation so far – a short summary. A short summary of what has been discussed here may be in its place: It is surprise-startle that triggers the affect-imaginative domain, and decides how we, with our irreversible perception, apprehend both *when-in-time backwards* and *when-in-time onwards* (the apprehension of *when-in-time backwards* we may circumvent by "subliminal stimulation", Chapter 6). If we limit ourselves to the affect in the affect-imaginative domain, it is in the affective shape of *feeling* that we "are" in the irreversible time, with death as the extreme leading edge of life (in the regressive state of deep depression it is the synchronous principle of time that reigns: "as it is now [in the suffering], it has always been and will always be in the future").

Before I dismiss completely the operation's importance for our experience of time, as "earth-bound beings", some additions to the above description of them. With Jean Piaget as a guide, an account follows of the emergence first of the concrete cognitively operating (six to seven years old) followed by the formal cognitively operating (eleven to twelve years old). A feature that ought to contribute to the difference between the two forms of operation may be involved in the development of language in the child. Before language has reached the level of "abstract notions" as an asset, the schoolchild is limited to having the material things within the visual field in order to be able to cognitively deal with them. In my fourth book in Swedish, there is a chapter with the heading "Intermission about the notion of 'notion'". That section, I can see in retrospect, may seem a bit obscure to the reader. My first clarification here I present as a question (in order to approach the problem in more detail in the next chapter): Would it not be worthwhile to have a distinction so that notions tied to "binary differences" could be assigned to one "class", and notions tied to "binary oppositions" to another "class"?

In the light of the evolutionary perspective, everything suggests that the most recent superstructure in Man is language. The cognitive operation, I've argued in earlier books in Swedish, probably precedes the emergence of language. The characteristic of the cognitive operation is "reversibility tied to the development of Decentring" (Piaget). It was probably through the establishment of abstract notions and recursion – created by Man for Man – in language that started the human being's staggering journey outwards to the macrocosm, and inwards to the microcosm. Two more questions: Is the operative cognitive development self-sufficient and independent of the development on the affect-imaginative line? Has the operative development come so far, all the way to the clever reflective abstraction without support from or bridging of the line of affect/internal presentation? But before I take on the task of trying to answer these questions, some further characteristics (that are hinted at in my two latest books in Swedish) worth conveying about the cognitive operation and time.

One attribute of the digital operation is that it is limited to two states, interfaces, – neutralizing of the clock time – to stabilize against, and also, if not endless then at least a "long strip" with room for many repeats of the two states, in order to fulfil its task. Simply put, the digital operation has one, and only one, fixed relation to time – *moments*. With the aid of the language of mathematics, Alan Turing succeeded in cracking the code for this relation between space and time (which he called "recursion"). In this way, we humans live today in space as well as in cyberspace. In the face of death, the digital comes out badly. At the extreme edge of life, the last interface remains – but on the other side of the border in death there are no other interfaces for the digital to stabilize against. This line of thought reminds us of what Anthony Wilden says: "Natural death is not therefore a part of the [digital] psychic system at all, and it is in no sense a principle of organization at the social level."[32]

Yearning: our most advanced motivational order

I have thus eliminated the digital operation, all the way to its complete form of reflecting abstraction, with regard to its influence over Man's sense of time as a living being on earth (besides measuring it), and over his awareness of death. This leads to the idea that it is probably only on the line of affect/internal presentation that features of importance to Man's awareness of death are to be found. As appears above, it may not be allocated to the activation of the immediate recognitive internal presentation in direct interplay with the immediate evocative internal presentation that defines perception (on the visual plane, structured around the age of three years). My suggestion implies that there are a number of necessary preconditions that need to be fulfilled during ontogenesis: (i) the emergence of evocation of episodes in time order back in time; (ii) the emergence of the ability for inner speech; (iii) the emergence of concrete cognitive operations – in all, established not until school age (six to seven years). But these are only the first steps towards structuralization that characterize us as adults – a structuralization that is established before the teens, at an age of eleven to twelve years. As a motivation, I'm inclined to attribute all this to a new form of interface, with a new motivational order, and this new emergence of interface, I repeat, is always digitally created. My proposal is that the advent of the cognitive operation (first concrete cognitive operations) is important for the formation of the interface. One usually speaks about "the power of thought" as being strong. That saying includes an often disregarded motivational order – in its dynamic constellation hinted at in my fifth book in Swedish. But let us take it all the way from the beginning, from bottom-up.

32 Anthony Wilden, 1972/1977, p. 144.

What (from the age of three months) is foremost triggered within the infant is the gradient or stimulation increase affects of surprise, fear and interest, and/or, as we have seen, from birth the level of the affect distress. Crucial for which of the gradient affects are triggered is the level of security/danger of the situation in which the child finds him- or herself. The strongly toxic or malign affects of anger, dissmell and/or disgust would be triggered in situations involving *traumatic* features. Depending on the extent to which the traumatization is present in the relations to the caregivers, or in specific situations during epigenesis, it may be important for the continuing development of the child as an affective and relational being.

Against the backdrop of my experience as a clinical researcher, clinician and psychotherapist, the level affect distress is guarded and "protected" during childhood by the immediately triggered gradient affect of fear. The latter may lead, together with the latticed shame as an included feature, to an inclination to anxiety-tension within the child – but with a certain tendency to breakthrough after adolescence and puberty. So it may abide for a whole (neurotic) life. But under the surface of the perpetual proneness to tension awaits distress, with its more or less pronounced inclination to breakthroughs in the shape of cry and/or depression. Distress is always present, to some extent, in the feeling of sadness – but also in the depression (the melancholy) as a more or less psychopathogenic state. To this I shall return in Chapter 6.

The specific openness and the flexibility within distress during the childhood, and continuously in our adult lives, ought to contribute to the fact that distress, as *secured* and *separate* scripts in *Srip* (as emotion), is rare, if at all existent. In *Srip*, distress is probably always an element in the script formation of fear and/or shame, the latter affect leaving the gate open for interest, as a feature of the script formation. But the feeling of sadness gives the possibility for extroversion; and with that, there is, at the same time, a possibility for a certain creative potential (see below). Or the gate of shame prevails, being shut and without access to interest – distressed directed inwards towards depression in different grades, and/or fear, and/or anger, and/or dissmell, and/or disgust. In the long run this probably makes for some type of psychopathological state (Chapter 6).

However, distress is the necessarily underlying affect for us as adults to be able to manage situations of loss and/or lack by more or less patience, and by the level affect distress transformed in sadness as feeling, by crying, tears and sobbing as expression/"language". Through distress as script formation by interest and enjoyment, there should in the long run exist extensions that can take the form of both *care* and *love*. This script formation extensively towards *mood* (Tomkins) may more or less obviously characterize the individual's personality.

Now, my proposal is that all that is decisive for the adult personality of the individual is the advanced *motivational yearning, a formation of an interface somewhere in the order of the System of Evocative Internal Presentation (Seip) on the one hand, and the System of Evocative Internal Images (Seii) on the other*. Yearning may be regarded as a superimposed motivational order on desire. More parenthetically, the central role of anguish in our grown-up lives makes Tomkins an adherent of the "tragic view of Man".

Yearning, with distress directed inwards, needs to have a stable support from interest, directed outwards, and from enjoyment as a discharge, in order to fashion *feeling* in the individual with extension towards: on the one hand inner security, equanimity and empathy in partnerships; and on the other hand, atonement all the way out towards becoming a moral, social and societal being. But the claim that distress is an absolute necessary feature of Man as Man, with feeling of sadness as guidance, implies that "irreversibility" is a quality of the flow (mood) of feeling, as well as of the flow of time. Affect/internal presentation comes no further than to evocative expressions – never, as is possible on the line of operation, to "reversible operative expressions". Feeling slavishly follows our immediate experience of time being irreversible. As a living human being I may feel sadness at times of loss – the affect of distress as our perpetual companion – the loss of a mother, grandmother, or father, grandfather … . My destiny is like theirs in the linear time to which we as "earthly beings" are subordinated. Not even the most cognitive operation can change that and convert it into something else, that is, convert death into an "abstraction". It is how we, within ourselves, come to terms with the affect of distress that constitutes one of the keys to what kind of person we become and are.

This may be transcribed in terms of "me-ness". "The me" has reached "me", as a responsible agent of my own feeling, my feeling is mine, as my death is mine. It is the feeling that can motivate and create openings for creativity, thus functioning as a driver onwards in time towards transcendence; for instance, feeling may be a help/necessity(?) for transforming the reversible reflecting abstraction toward "discovery"; or as within the artist in the form of a figurative and/or linguistic "concreteness". It is the reversibility within operative cognition that makes it possible to turn around all the way back to search for – *Erklären/explanation*. The immediate feeling never gives any "explanations" – however, in the most favourable of instant/ Augenblick feeling may enable *"Verstehen/interpretation"* in the form of feeling as a clear derivation – that sometimes expresses itself in the form of an "aha experience". If we only take the state of "clarifying atonement" into consideration, dynamic insight therapy might be a way of reaching it. Here, only very briefly: it is *recursivity* and semantics within language that gives the "therapeutic dialogue" the possibility of returning to *when-in-time backwards*, when, at its best, it may reach atonement with the traumatization of the affect, and/or emotion, and/or feeling in the contingency S-R-SE.

Is there an ultimate common denominator for all psychopathologies? The human being can never abjure (ward off completely?) the negative affects. They are there from the beginning and continuously throughout life, more or less in the presence/present. It is probably not wrong to describe them in terms of being a perpetual "transitional space", between life and death, in the face of the presence/the now state in life as a repeated reminder of the inescapable death. This is in fact the charter for all shapes and expressions of psychopathologies. And ultimately it is about how we have come to terms with the script interest/shame/distress/enjoyment as unavoidable elements in the directed motive of yearning.

It is with the onefold grounded affect within the human being transformed into feeling as it is with the irreversible time for Man on earth – it never turns around to see if some errors have been committed and if the error may be corrected, if one starts all over again. One of the possibilities for doing that, within the human being, resides in language. That I leave to the following chapter to discuss.

CHAPTER 5

Homo psychicus: a brief comment on language

> In science, we are of necessity dependent on language. It is semantics in language that both can save psychology and lead it all the way down to the origin of language, to the layers where the inborn and biological given affects reside. Each linguistic sign's tacit message, used by everyone, is to disregard the sign as a symbolic sign.[1]

Language and origin

Nobody knows for sure when language, with its semantics and syntax, emerged during phylogenesis. There have been many hypotheses. But linguistic researchers with an evolutionary orientation agree on one thing: language did not emerge overnight. Here follow some of the proposals.

If we follow Michael Corballis, this applies:

> With a primitive grammar, consisting only of symbols for objects and actions, combinable into two-word sentences, generative language is born. [...] [A] more sophisticated language probably did emerge over the past 2 million years and was accompanied by more innovative and generative thought processes that enabled the development of more sophisticated tools when circumstances demanded it. However, language was probably primarily gestural rather than vocal, so the development of manufacture may have actually been *inhibited* by the use of the hands for communication. This could be why manufacture did not really begin to flower until speech became the dominant mode, perhaps as recently as 50,000 years ago, thereby freeing the hands.[2]

1 Alf Nilsson, 2018.
2 Michael Corballis, 2002, p. 37, pp. 88–89.

Nicholas Humphrey considers the first cave paintings showing humans in the Lascaux caves, which may be dated back 17,000 years. According to his speculation, this ought to be date for the establishment of the first language with semantics and syntax.

> Maybe, in the end, the loss of naturalistic painting was the price that had to be paid for the coming of poetry. Human beings could have Chauvet or *The Epic of Gilgamesh*, but they could not have both.[3]

With reference to C. Henshilwood and T. d'Errico, Teresa Bejarano argues that the finding in the Blombos Cave in South Africa of an engraved stone with series of lines, one below the other, points to a spoken language with syntax and grammar that may be dated back 60,000 years. The revealing indicia, according to Henshilwood and d'Errico, consist of the fact that one of the series was finished before the beginning of next. On the basis of this, one may read that the engraving initiated "paying attention to the final whole, [...] a last step". It presupposes "the great requirement of the numeric ability", thus, an inner ability to comprehend an ordered coherence before-after. Because without an inner comprehension of order: 1, 2, 3, ... *n* – always accompanied, eventually, with an understanding of the possibility of $n + 1$ independent of the number – that is, unless "the series of lines in one direction were completed before the series of lines in the other direction was begun", there would have been no ability for syntax".[4]

In an earlier article I have considered in detail Dean Falk's theory (see also Chapter 2) of the emergence of language from a phylogenetic perspective.[5] During the hunting and war raids of *Homo erectus* the mothers and the children stayed put at a more or less permanent place, where they collected edible fruits, berries and nourishing root vegetables. At this point baby slings appeared – according to Falk, one of the oldest man-made tools (dating back 1.6 million years). From time to time, though, a mother needs to put down the child on the ground in order to get at less accessible root

3 Nicholas Humphrey, 1998, p. 166.
4 C. Henshilwood and T. d'Errico, 2002/2011; Teresa Bejarano, 2011, p. 294.
5 Alf Nilsson, 2010.

vegetables *("putting the baby down")*. That triggered an alarm call in the child in the form of screaming/crying. But as I said, losing bodily contact with the mother ought to trigger the alarm call "danger of lack of attachment relationship". Within the mother, infected by the child's distress, a first sort of *feeling* of care was probably triggered, resulting in lifting up and rocking the child, in order to comfort him/her with "lullabies". This song is believed to have had:

> the same musical features as *Motherese, the adult-to-child language*. According to Falk, the mother's song started with humming and simple melodies that eventually developed into lullabies. Everywhere on Earth, "mother's songs" appease small children. [...] Of all the speculations about the origin of language, this is the most beautiful: Language arose from crying, comforting, and singing.[6]

As a reader, you are already acquainted with how Teresa Bejarano, when deriving syntax, starts from the newborn child's "expectation a priori". To that, expectation a posteriori is added. Expectation a priori, as we saw above, is present as a feature of the inborn reflexes; expectation a posteriori, as a contribution, is unique for Man. Generally, as is the premise of this book, expectation does not wait for any specific object/goal, but waits for something within certain limits: a big enough ball in the hand for expectation a priori to grip ...; or, a kinaesthetic action of some other person equipped with a physical object/manufactured thing/artefact for expectation a posteriori to mimic – mimicking, in the continued development is converted into imitation. This makes for a sensorimotor operation, functioning as a precursor to the operative cognition's creation of differences between *classes* of object that we try to bridge with *linguistic notions*.

In line with Michael Tomasello's opinion is Bejarano's basic thesis that syntax in language is not as much due to biological evolution, as to human history as "memetic", or to the human being as a social and cultural being.[7] This chapter ends with a very brief summary of Bejarano's derivation of syntactic elements and their way to full grammatical syntax.

6 Alf Nilsson 2010; Dean Falk, 2009; Jan Einarsson, 2013, p. 26.
7 Michael Tomasello, 2008; Teresa Bejarano, 2011.

Language and ontogenesis

There is no given knowledge, either, about how the child learns the mother tongue from the beginning. In a recently published book, Steven Pinker writes:

> Despite over a decade and a half of vigorous debates, we still do not know what sort of a priori knowledge, if any, is necessary to learn a natural language; [...] nor how semantic information affects the learning of the syntax of a language.[8]

Patricia Kuhl has dedicated a large portion of her research to the study of children's responses, spoken language from birth: "At birth, infants discern differences between all the phonetic units used in the world's languages". Kuhl's studies show that the newborn child has a readiness to intercept and incorporate any of the more than 6,000 of the existing world languages. J. Mehler *et al.* have shown that French infants, four days after birth, suck on the pacifier more intensively when listening to French than when listening to Russian. T. Nazzi *et al.* show in a study that French infants of the same age express different bodily patterns depending on whether they listen to sentences in English or sentences in Japanese. At the age of two months, children filter – put up constraints on, see below – other languages than the mother tongue. The conclusion one may draw from these results is that it is only during an extremely short period that the child has an unlimited readiness to acquire all phonetic expressions and languages in the world. To this one may probably add Pinker's opinion (see above): "The child has no manual to consult, and presumably that is why he or she needs innate *constraints* [for language ability to emerge]" (italics mine). Pinker often returns to "neuronal developmental constraints" in his book – worthwhile considering with reference to "the epistemology" presented in Chapter 2 (Terrence Deacon).[9]

8 Steven Pinker, 2013, p. 5.

9 Patricia Kuhl, 2000, p. 101; J. Mehler, G. Jusczyk, G. Lambert, N. Halsted, J. Bertoncini and C. Amiel-Tison, 1988; T. Nazzi, J. Bertoncini and J. Mehler, 1998; Steven Pinker, 2013, p. 136.

Language

In my first book on *Homo psychicus* in Swedish, there is a lengthy chapter on language with subheadings such as: "The structural building of language", "Life and communication – animal and vocalization – chimpanzees and symbols", "Language and evolution – language and brain – language and anatomy", "Language and the child's epigenetic development". In the books that followed clarifications, sometimes modifications and even changes of the terminology have been successively made. To this, other additions adapted to the course of time have been inserted. Below you can find elements taken from earlier editions, along with some scattered thoughts in order to convey something about my present view of language.

I start with a long quotation from a recently published book, *Språket och språkarna* ("Language and language users") by Jan Einarsson. The purpose of the quotation is partly to refer to the content presented in the introduction to the chapter, partly to introduce the reader to the content that follows the quotation:

> The social function of language is closely associated with an emotive function, expressing feelings of closeness and distance. Adults in many different cultures talk to small children in a way that creates special intimacy. In English this is called *Baby Talk* or *Motherese*. In Swedish we have used the term *vuxentillbarnspråk* "adult-to-child talk". The central message in such talk is conveyed by the emotional (cutesy-cute) tone and not necessarily by the content of what the words actually say. The fact that this form of communication is universal and mostly wordless means that one may assume that *there was a fundamental emotive function in man's early speech. The informative function of language presumably developed much later than the emotive function.* (latter italics mine)[10]

10 Jan Einarsson, 2013, p. 25.

Steven Pinker asks three questions that guide my text that follows later on:

> What is the function (if any) of language? What are the engineering demands on a system that must carry out such a function? And are the mechanisms of language tailored to meet those demands?[11]

As appears in my first book (and in the succeeding books) in Swedish, I describe the affect function in terms of being the "first *Homo* language" or, the first "engineering demand" of *Homo* beings directed towards *mutual communication* (affect communication within mammals beyond the human being is solely based on unilateral sequential patterns).

Einarsson's suggestion about the position and importance of emotions/affects in language coincides well with mine. That proposal may hardly have escaped you as a reader; it is also the pervading theme of this book. Moreover, different levels of development on the line of affect/internal presentation may, through emotion and all the way to feeling as an expression, adopt the quality of mutual communication with differing depth and breadth, ultimately transformed into human love.

I take Silvan Tomkins's proposal about the *abstractedness* of affect as one of its specific characteristics as the starting point when pointing to *one* possible transition to language. It is within the affect of *anger* that the abstractedness is most undisguised. I will now let Tomkins argue for my cause. This will be done through the rendering of a long excerpt, not quoted before in any of my books in Swedish:

> Affect amplifies, in an *abstract* way, any stimulus which evokes it or any response which it may recruit and prompt, be the response cognitive or motoric. Thus, an angry response usually has the abstract quality of the high-level neural firing of anger, no matter what its more specific qualities in speech or action [in my view, that there would always be an element of "transformed abstract anger" in speech and action]. [...] The appropriate minimal paradigm here is the miserable, crying neonate who neither knows why he is crying nor *that* there is anything to do about it; nor, if he thinks that something might be done, does he know *what* he can do about it. It is *not* a "motive" in the sense in which psychology has used that concept, though it

11 Steven Pinker, 2013, p. 121.

> provides the core of what may become a motive. Without affect amplification nothing else matters, and with its amplification anything *can* matter. (brackets mine)[12]

First a clarification of an aspect in the quotation that we already know the meaning of, namely that the amplification of affect involves *both* stimulus and response. Amplification is "abstract" in the sense that the response is never predetermined, neither in kind nor in degree, thus there is no final causal relation between amplification of stimulus and response (*Homo psychicus* is and has always been an innovative being).

I begin where the quotation from Tomkins ends – the point that I chose as my starting point: the birth cry of the newborn child. I remind you that the prematurely born child has parts of the foetal development to catch up. Moreover, my suggestion is that the level affects of anger and distress are indistinguishable from each other, at the time of birth – and thus, "birth scream/cry". It is not until the age of three months that the child cries with tears. It is also the age when the caregivers (the mother, as has been shown, is a step ahead of the father) make a distinction between crying expressing lack rooted in distress from cries expressing anger (see above).

Before I return to the theme of the abstract affect, which could be a feature of language, a small interlude on the theme of anger and distress, and their way to the contingency of S-R-SE (Someone in Relationship to Someone Else), and further to *Homo psychicus* as a social and societal being. Once more, I will let Tomkins be my spokesman:

> All the negative affects trouble human beings deeply. Indeed, they have evolved just to amplify and deepen suffering and to add insult in the injuries of the human condition in several different ways. Terror speaks to the threat of death to life. *Distress* [italics mine] is the affect of suffering, making the world a vale of tears. Shame is the affect of indignity, of defeat, of transgression, and of alienation, striking deep into the heart of the human being and felt as an inner torment, a sickness of the soul. But *anger* [the most abstract of the affects, italics mine] is problematic above all other negative affects for its social consequences. My terror, my distress, and my shame are first of all my problems. They *need* never become your problems, though they may be. But my anger, and especially my rage, threatens violence for you, your family, your friends, and above all for our society. Of all the negative affects it is the

12 Silvan Tomkins, 1991, pp. 69–70.

> least likely to remain under the skin of the one who feels it, and so it is just that affect all societies try hardest to contain within that envelope under the skin or to deflect toward deviants within the society and toward barbarians without.[13]

In the phylogenetic perspective, probably prevailing when anger and distress were established as separate affects, they pushed forward two sides within *Homo psychicus*: without distress and sadness (as feelings), no potential for care and love as feelings; without anger, no potential for conflicts, animosities, warlike acting out with manslaughter as a result … .

But now time to return to language and its origins. I have already shown that surprise, with certain traits of either-or, is an affect that always supports expectation directed towards that which exists – but also towards that which does *not* exist. Tomkins has no doubt that anger is the most abstract affect among the given affects. Can it be that it is the negative affect of anger that is the basic feature of the penetrating, pervasive, and rampant abstract cognitive operation? And that it is the affect anger that lies behind the fact that some *Homo* being had an experience involving the senses, at some point in time, that founded the (upcoming) words/signs *what*, *where*, *when* and *how*? In the beginning all in one paragraph, and eventually it was decomposed into: What's up? Where is it taking place? When to strike? And how, in what way? Parenthetically so far: it is probably much later during phylogenesis that the content of the word/sign *why* found its way within the *Homo* mind. And by the way, Aristotle probably knew this. About anger, he writes (see Chapter 3) as follows:

> Anybody can become angry – that is easy; but to be angry with the right person and to the right amount, and at the right time, and for the right purpose, and in the right way – this is not within everybody's power and is not easy.[14]

If we are ready to follow this line of thinking, anger is probably the source, with its highly abstract features, from which the operative line of development retrieved one of its permanent borrowed elements that successively led all the way towards the capacity of cognitive operation for abstract

13 Silvan Tomkins, 1991, p. 111.
14 Aristotle, 1893, Online Library of Liberty, 2.9.

thinking, ultimately for reflecting abstraction. But let us return to the beginning, to the *Homo* being equipped with: outwards-in-inwards as a circular orbit; eventually followed (phylogenetically)/follows (ontogenetically) the operatively directed *either* outwards-in *or* inwards-out; eventually followed one of the most simple operative abstractions *either* outwards *or* inwards. If this proposal is valid to any extent, thus, affect could be one of the "innovators" of the abstractness of operative cognition, ultimately equipped with the capacity to abstract macrocosm as well as microcosm. But let us not forget – it is language that ultimately made Man into this miraculous cosmologically outward- and inward-directed being.

Here a small reminder about what I already touched upon in Chapter 3. The physical thing/manufactured object/artefact one may have *or* not have, touch/pick up *or* put down/put aside, at the same time as its usefulness remains intact. It is the usefulness that decides its "value", *either* in use, or *stored* still being useful (when an opportunity is given). In relation to living persons (from the beginning the caregivers) – relationships are always based in the affects – the relationship ought to be totally different: One cannot "pick up", and "put down/aside/away" for use. Their "value" consists in, on a continuum with the poles present and absent at either end (see books 4 and 5 in Swedish for my discussion in detail), permanently functioning as mirroring of affects/emotions/feelings to us.

Having come this far, I need to admit that I've not got any further with the questions I asked in Chapter 3, which are repeated here: When it comes to the human world ("institutional facts") – with *binary oppositions* as the ultimate device for dealing with the world – why have we not demarcated and developed a specific "notion code" for it? And this considered in relation to the world of physical things/manufactured objects/artefacts ("brute facts") – with *binary differences* as the ultimate device for dealing with that world? Is it this world, and only this world, in which "abstract concepts" are "valid"? Is it affects/emotions/feelings – creating continuity and coherence of "me-ness" – and as involved in "perception", as perception is defined in this book, that make up the constraint? Or is in the "transitional space" of "institutional facts" that notions of "ideology", "politics", "morals", "belief" etc. emerge, forming the constraint that ultimately prevails?

To round off the answers to Pinker's initial two questions about the "structure" and "function" of language, I will reiterate my previous proposal: The human being's affective relations to persons, according to the contingency S-R-SE, ought to be the foundation for semantics – the meaning/understanding – in language; the operative relations of Man to the useful thing/object/artefact ought to be the foundation for the syntax/grammar of language.

With that, Pinker's third question remains, about whether the mechanisms of language are tailor-made for an interjective, communicative and informative language. Immediately when I read the question a counter-question arose in me: Where to seek and find this creature, this being of cosmic dimension who may answer that question? Hardly here on earth. But a simple answer may be that language is well suited for the subtlest of poetry at one end, and at the other end, for rendering the most intricate of conditions on microcosmic as well as on macrocosmic levels – mathematics. But this does not rule out that I will take some extra rounds with the issue and give it further consideration.

As has been made clear, the starting point is that language is a "tailored" superimposition on the human being as an affective-internal-presentational and cognitively operative being. Is it as such, an optimal human formation, according to Pinker himself, in order to: (i) "acquire [...] information about the world", (ii) "by tapping into the vast reservoir of knowledge accumulated by other individuals", (iii) "avoid having to duplicate the possibly time-consuming and dangerous trial and error process that won the knowledge", (iv) "within a group of interdependent, cooperating individuals [experience] the states of other individuals".[15] Against the backdrop of this account, I cannot think of anything that language could not convey. But in order to be able to compare, I cite what Einarsson thinks language can be used for:

1. To express the speaker's state (*symptomatic* function), sometimes described as the *emotive* function
2. To influence the listener (*signalling* function)
3. To describe the world (*symbolic* function)

15 Steven Pinker, 2013, p. 122.

Einarsson adds to these three uses a further three, denominated by Roman Jacobson:

4. To establish contact (*phatic* function)
5. To describe language (*metalinguistic* function)
6. To focus on the form of the message (*poetic* function).[16]

There are obvious overlaps between Pinker's and Einarsson's descriptions of what language is "tailored" for. A certain difference is that Pinker primarily underlines what language can be used for epistemologically, while Einarsson stresses the more indirect use in the interaction between human beings influencing each other. As should already be clear, Einarsson holds the view that the original function of language was emotive rather than informative, the latter emerging later during phylogenesis, according to him.

For you as a reader, there ought to be no doubts about my following in Einarsson's footsteps. In the fifth book in Swedish I explain how Teresa Bejarano follows the same line of reasoning with her derivation: connotation/intension → denotation/extension → syntax. Is there anything to add to this? Relating to Charles Sanders Peirce's parlance, I argue that the icon is in its foundation a formation on the line of affect/internal presentation embedded in a vocalization that gives it value and meaning. This is in line with Anthony Wilden's thinking: "The iconic sign is an instance in the emotional."[17]

So far, you have probably apprehended that my proposal about language and its emergence rests on the idea that language is an example of a *specific emergence* during evolution and the course of phylogenesis. Are there other opinions and theories on that? It is well-known that Noam Chomsky early on argued that language is such a unique contribution to Man, and that it lies beyond the scope of evolution/phylogenesis. Language is such a complex and specific formation that it must have emanated from an unlikely rare *mutation*. What Chomsky really meant was syntax in language.[18] To

16 Jan Einarsson, 2013, p. 97.
17 Anthony Wilden, 1972/1977, p. 270.
18 Noam Chomsky, 1957.

that may be added that Chomsky, during the years that followed, modified his previous standpoint and attributed evolution/phylogenesis prerequisites for advanced accomplishments commensurate with language. The following is a quotation from Pinker, taken from Chomsky:

> In this regard [*the evolution of infinite digital systems* (Pinker's comment)], speculations about natural selection are no more plausible than many others; perhaps these are simply emergent physical properties of a brain that reaches a certain level of complexity under the specific conditions of human evolution. (italics mine)[19]

My own thoughts on this are, that if language – semantics and syntax included – is an invention of evolution/phylogenesis, it requires an evolutionary/phylogenetic invention of the digital operative line of development. Firstly, a general viewpoint on this, later on in the section on syntax/grammar I will try to formulate myself more precisely. Considering all the discoveries about earth/the world/universe/living creatures (evolution) that the human being has made, there must be something in him or her, which in its turn makes it possible to transform *this within him or her* to what exists outside him/her. Schematically it probably has to do with a change of quantity – serial formation of neurons in the brain – in combination with quality – cyclic circular neuron/transmitter formations in the shape of networks in the brain.

Stephen Jay Gould launched, as a warning for "naïve adaptationism", the idea that the so-called *"Spandrel proposition"* could be a possible source of the emergence of different phenomena and constructions, particularly language, during the course of evolution. The theory is based on an analogy between the human world and engineering: When constructing bridges that rest on pillars, one gets arch bridges as a result. These arches do not have any function – but when in place they may get new specific functions, as practical passages under the bridge and more or less decorative, attractive bridge arches and the like. This implies generically that the emergence of evolutionary constructions without functions during a certain phase of evolution may in a later stage have become a distinctly functional/structural contribution to the construction of life. Pinker dismisses "the theory" claiming: "that *natural selection is the only scientific explanation of adaptive*

19 Steven Pinker, 2013, p. 139; Noam Chomsky, 1988, p. 22.

complexity".[20] It is noteworthy that Gould's suggestion, and also similar proposals of other authors, is based on a staggering analogy retrieved from advanced structural engineering in the human world.

To close, a few words on syntax/grammar as a feature and control instrument of language. It is syntax that, through *recursion*, ensure that the right word ends up in the right position, at the right time in the serial flow of the spoken language, as it does in the late innovation of the written language 5,000 years ago (despite the fact that it is considered to be the "cradle of civilization", I take that no further). My overall proposal, presented in detail in previous books, is that without the digital *reversibility* of cognition (retaken from within-out and from outside-in in one and the same moment), there can be no recursion, the latter being characteristic of syntax (together with, according to Einarsson, "structuring", "hierarchization" and "flexibility"). Wilden's proposal is that: "Words are digitalization of time ...".[21] I'm inclined to reverse Wilden's proposal: It is (as we know so far) in nature, at the level of elementary particles, that space has "digitalized time" as "a square in space" (as Man has achieved within the digital machine) and which lies behind the twofold on the operative line of development in space and time on earth (and naturally, Albert Einstein's formula $E = mc^2$, describing space-time in the cosmos, comes into play). As an extension of this, the emergence of syntax during the phylogenesis is no longer a mystery to me.

Being neither a linguist nor a psycholinguist, I find what Einarsson denominates as "the double articulation" harder to understand in terms of being a result of evolution/phylogenesis: "The [double articulation] implies that in principle there is no connection between the auditory or written configurations and their meaning. The relation between form and content is arbitrary, that is, random."[22] When it comes to understanding this, linguists and semioticians probably have better suggestions than I have. This means that I refer you to them to answer the question. But as regards the statement that "Grammar is a notoriously bad choice of medium

20 Steven Pinker, 2013, p. 115.

21 Anthony Wilden, 1972/1977, p. 375.

22 Jan Einarsson, 2013, p. 78.

for revealing emotional patterns" I take Einarsson outright at his word.[23] The last commentary so far will become a hope that the reader learned a bit more about the fundamental importance the affects probably have for the emergence of *Homo psychicus* as an affective-internal-presentation-imaginative, operative, but also as a linguistic being.

Mathematics as language

Einarsson notes another connection that points to both language and mathematics as "language":

> The link between speaking *tal* and counting *tal* is shown by the fact that words for *tal* "talk" and *tal* "something to count with", "number", "quantity" obviously have the same origin, the same root. In Swedish we have the clearly related words *tälja* "carve in wood", with the archaic sense of "count, tally" and *(for)tälja* "recount, tell".[24]

Above, Teresa Bejarano's opinion that in the ontogenetic development the child needs to be able to handle cardinal numbers as a prerequisite for the development of an ordered speech, is made clear.

We have serial and continuous visual perception in the form of Gestalts (right hemisphere), but we do not think of it very much. To this we may add serial talking/speaking, through hearing, with the right word in the right position ... which comes about with the staggering speed of a moment (left hemisphere). Regarding the human access to multimodality, one may understand how the quick visual perception in Gestalts might have "infected" the hearing with an instantaneous vocalization of words/sentences. This may be put up against Einarsson's claim regarding the chimpanzees:

> The difference between a chimpanzee and a human being implies, for instance, that the chimpanzee brain is directed to process inputs from sense, vision and hearing

23 Jan Einarsson, 2013, p, 127.
24 Jan Einarsson, 2013, p. 54.

> *separately* [and thus not in a multimodal way, according to my interpretation of Einarsson he refers to the evocative level]. (brackets mine)[25]

To sum up, the abstract feature of primarily the affect of anger, as response (with the triggered affect response always being undetermined when it comes to kind and degree) together with the fundamental affects for determining value and meaning, we have the foundation for understanding how the almost 7,000 languages in the world may both differ from each other, and be similar to each other. It may perhaps give an idea of how to understand the human hybrid that language is: both in terms of the digital syntax that separates languages from each other, and the analogous semantics that gives similarity in terms of meaning and value. The latter makes it possible in principle to translate all languages.

Jonathan and languaging

The last time the reader heard anything about Jonathan, he had in fact already turned two years and eleven months. Now I turn the clock back to when Jonathan was exactly seventeen months old. He was still a one-word speaker, although sometimes with a tendency to two-word expressions in situations where he pointed at the listener while simultaneously guiding her/him. Jonathan at seventeen months runs around steadily on his legs – being very lively and full of energy. He is always curious about what might be hidden in closed drawers and cupboards in front of which he often stops, and then he opens them and takes out things that mostly, after a quick examination, end up on the floor. This is also the case with toys, spread over the floor: he may stop in front of some toy, pick it up and watch it for a couple of seconds – soft things he draws to his cheeks, hard things he watches – and drops them back onto the floor and then moves on. It has been like this since he started his multiple-step walk. One thought: That Jonathan hardly stops for "contemplation" – maybe that has to do with the fact that surprise/expectation not yet has established itself in the

25 Jan Einarsson, 2013, p. 22.

positive and negative gradient affects as emotive scripts? Moreover: is it the case that, to be able to "contemplate" requires that Jonathan to have reached at least a first embryonic formation of "targeted motive"? During the work on my fifth book in Swedish I paid extra attention to the development of Jonathan's "patience".

As I have already touched upon, Jonathan had at that time demonstrated a few new skills. The first time he pointed, he was fourteen months and seventeen days old. His mother and father had been putting the lasagne in the oven. After the meal, his mother happened to say: "Where is the lasagne, Jonathan?" He then ran to the oven in the kitchen, pointed at the oven door with a "Da!" and directly afterwards gazed happily at his mother's face and eyes. When his mother and father repeatedly asked "Where is the lasagne?" he moved towards the oven, pointed at it with an expression of enjoyment along with the vocalization "Da!" – and thereafter he gazed quickly into his mother's/father's face and eyes ("Well done Jonathan!"). From that day on, Jonathan had seized index finger pointing, and pointed, when requested, at things here and there – with a perpetual extension of physical things/manufactured objects/artefacts to be included in his pointing register.

When Jonathan visited his grandmother and grandfather, at the age of fifteen months and nine days, and being with his grandfather and mother in the kitchen his grandfather put the same question to Jonathan: "Where is the lasagne?" Immediately Jonathan went towards the oven and pointed at the oven door with a loud "Daa!" – and immediately after his grandfather, in return, met Jonathan's gaze of enjoyment directed towards his face and eyes. After a while, his mother repeated the same question – and again Jonathan hurried towards the oven and pointed at the oven door.

The conclusion I'm prepared to draw from this, without hesitation, is that Jonathan was on his way to coming to grips with "categorizing similarities (oven door as oven door), linked to recognition via linguistic expressions". On a generic level, the conclusion is that the little child has an extended language *recognition* which is far more advanced than the ability for language *evocation* in ontogenetic development.

At the same time, Jonathan, aged fifteen months and eight days, is offered a segment of an orange that his mother just has been peeling. After he has eaten the piece of orange, his mother hands over a bit of the peel to him which he grabs in his hand, at the same time as his mother asks him to "go and throw the peel away". Jonathan goes directly to the cupboard where the bags for garbage disposal are, and throws the peel in the bag for food waste. It is obvious that Jonathan at this age can understand a verbal request and transform it into a perfectly executed action sequence. My conclusion is that without access to both *intention* and *imitation* he would never have been able to do this.

In my first book on *Homo psychicus* there is a detailed account of how chimpanzees and bonobos cope with the handling of "lexigrams", and with the performances they are capable of when it comes to linguistic requests. If one is acquainted with comparative linguistic research, one knows who the bonobo and celebrity Kanzi is – the primate researcher Sue Savage-Rumbaugh's particular favourite. On one occasion, when Kanzi was in a pool, he was given the task and the question: "Can Kanzi get some grapefruits and throw them into the pool?" He immediately got out of the pool, and went (knuckle-walking) to a table with a bowl of grapefruits, not far from the pool. Without effort, he picked some grapefruits and threw them into the pool. Savage-Rumbaugh claims that Kanzi has a "linguistic capacity" comparable to that of children of an age of two-and-a-half to three years. But, no one ever heard any word coming out of Kanzi's mouth.[26]

Is there anything, so far, to add to Jonathan as a languaging little person? Besides expressions of single words, as correct descriptions of different objects, he has a number of personal vocalizations. When he points at different things, the level of eagerness/interest seems to have an importance for whether the vocalization is going to be "Da", "Da, da", "Daaa", Da, da, da" etc. During the Christmas and New Year holidays, when Jonathan was sixteen months and ten days, with grandmother and grandfather visiting, there was a good deal of singing and play with Jonathan in mother's lap: "Grandmother's little crow" who slid here and there and then down into the ditch – to the great joy and laughter of Jonathan. A short exclamation of "Hi!" was his vocalization of "Once more!" When his mother

26 S. Savage-Rumbaugh and R. Lewin, 1994.

sang "Twinkle twinkle little star", he ran out into the kitchen and pointed, with the accompaniment of an eager, short and vigorous "Da!!!", at the poinsettia (the Swedish name means "Christmas star") that hung in the kitchen window. He had received a book with songs, pictures and text as a Christmas present. Among the songs in the book were "Grandmother's little crow" and "Twinkle twinkle little star". On the page with the song there was a little button to push, and when it was pushed the melody was played. During the time they were playing, Jonathan remained totally calm and still where he sat in the sofa, everything in a whole unity/one whole moment/synchronous/diachronous of: the listening, the affect assemblage, the tone, the rhythm, the Gestalt as instantaneous as well as serial … . Is Jonathan finding his ways in: Affect as emotion? Perception as diachronous? Cognition as the first primary operation? Language as an evocation of "everything in harmony in the moment/synchronous/diachronous …?" Anyway, it was something in the wordless melody that captured him, that aroused his interest extending to the remaining interest as emotional state. I remind you of Tomkins's opinion about music as being the closest in analogy with affect within Man. In Jonathan's expression of seriousness, I can see him being on his way to the patience of make-believe play.

A week later (at age sixteen months and sixteen days) Jonathan sits in the sofa in the living room. He follows his big brother Isak (eight years old) with his eyes. Isak got a remote-controlled helicopter as a Christmas gift. He keeps it flying, moving it up and down, within of Jonathan's visual field. Again Jonathan sits totally still in the sofa, his eyes following the movements of the helicopter, with exactly the same serious facial expression as I described above. Jean Piaget observes how children of that age are preoccupied with differentiating physical things/manufactured objects/artefacts that "do *not* move from within themselves" from "living persons who *move* from within themselves".[27] As a contribution to that, my suggestion is that the child on his or her way to transforming affects (primarily the affect script interest/shame/enjoyment) into emotive scripts in *Srip* as a kind of first "motive to follow" – with an intended listening as well as

27 Jean Piaget, 1951/1972.

with an intended gaze (now it is time for the notion of "attention") – but not yet with developed *targeted motives* (desire, and the next step yearning) that the child may follow fully towards acting, primary cognitive thinking, languaging. According to my proposal, it ought to be the case that the serial and continuous gazing "infected" serial listening, which eventually leads all the way to serial languaging. In Chapter 7, I will come to the question of how this can be.

* * *

The following section was written as a conclusion in my fifth book in Swedish. Jonathan was at the time nineteen months and twenty-one days old. When he was seventeen months and eight days old he had sat up in the bed when he woke, and repeated for around twenty minutes: "mammy", "daddy", "Isak" (big brother), "Ruff" (the family dog), "Laila" (grandmother), and "grandfather", not fully articulated. (So, "Ruff" sounded like "uff", for instance.) At the time when I was writing (Jonathan now being nineteen months and twenty-one days old), he used pointing only in exceptional cases. At the mention of things that were familiar to him (most often toys) he went straight away and fetched them. His mother and/or father often played DVDs with nursery rhymes for Jonathan. Among the favourites were Lennart Hellsing's "Krakel Spectacle" (*Krakel Spektakel*), "Mr Cucumber" (*Herr Gurka*), and many others. Jonathan stood up, swaying in time to the rhythm of the songs, sometimes with a type of singing sound: "lalala". But still at this point in time he did not express himself with two-word sentences.

A very brief review of Teresa Bejarano's derivation of syntactic links, predictive, theme-rheme and full grammatical syntax

The origins of syntactic links, predictive, theme-rheme and syntax "would certainly be linked to communication and interpersonality". In such communication and interpersonality, voice "joins the communicative force of *emotional* intonation". Verbs "would have function as a causal bridge, and their associated expectations of questions. [...] But if we admit that these words (those which would eventually become verbs) at first did not have any meaning other than that of expressing frustration [fear-terror] ['It's a snake!'], or relief [enjoyment-joy] ['It has gone!'] or surprise[-startle], then we will have to explain why they were learnt articulatory-phonetic signs and not mere *innate emotional* signals." Because "articulatory-phonetic signs gradually became more common and less costly, they would come to be used with the mere *emotional* discharge function." The expressive uses of a word "would be the seed of properly grammatical or abstract links [...] and subsequently also the crystallised grammatical links would be the seed of the ability to remember out-of-date perceptions." As regards ontogenetic and historic origins Bejarano proposes: "In children, abstraction from real links might be capable of reaching syntactic categories ... [...] My suggestion as regards historic origins will be based on *emotional* discharge speech." Let me finish this very brief summary of Bejarano's stance with one more quotation: "The expressive-function speech may sometimes be external, sometimes internal. This alternation will occur not only in children at the transitional stage from six to seven years of age (age of reaching internal speech), but also in adults. [...] [E]xpressive speech would be first external and then would be internalised. [...] The only thing that matters to us at this point is that we are dealing with expressive speech or an *emotional* discharge. [...] As regards expressive speech, it is clear that, for language to be aligned with phylogenetically ancient ways of expressing *emotions*, the habit of some linguistic elements has to be deep-rooted. On the other hand, meaning, with its links, would be the culmination of linguistic learning" (italics and brackets mine).[28]

28 Teresa Bejarano, 2011, pp. 359, 360, 312, 297, 271, 290.

CHAPTER 6

Psychopathology as viewed in epigenetical perspective

> My forgetting names and faces would be covered by *DSM-5* "minor neurocognitive disorder". My worries and sadness were going to be "mixed anxiety/depressive disorder". The grief I felt when my wife died was "major depressive disorder". My well-known hyperactivity and distractibility were clear signs of "adult attention deficit disorder" [ADHD] [...].[1]

Introduction

This chapter I will introduce with a description of psychiatry in Sweden. As an example, it is probably representative of psychiatry in the rest of the Western world. In Sweden, psychiatry is traditionally a speciality within healthcare. Psychiatry is thus a medical speciality. As such it has as its task to treat what once was labelled with ordinary denominations like "insanity" and "nervous troubles". These types of problems have throughout Modernity been associated with shame, and they still are. The philosopher and historian of ideas, Michel Foucault, is a modern author who has profoundly analysed "psychic, social and sexual deviations" in the perspective of Western societies during what he calls "the classic (modern) era".[2]

In the perspective of the affect of shame as being necessary for the child's socialization and society's civilization process, the shaming of the

1 Allen Francis, 2013/2014.

2 Michel Foucault, 1976–1984/2002, 1961–1972/2010.

"strange", "odd" and "deviating" within Man as a societal being is a natural component. The affect of shame is the one that is the most obvious. Here I would like to remind you that shame is the gatekeeper between the positive and the other negative affects. Among all the affects shame is the most contagious – for us to be attentive to the slightest element of the "strange" within the "self", but also within the "other". Because if it is about shame, then it is not about the malign fear – with threat and danger directed towards ourselves, ultimately towards extinction.[3] To demystify shame in front of "the psychic strange" is something of a myth the modern society cultivates, in the extension of which mental ill health is considered to be like any other illness. Shame – ultimately toxic – probably functions both like a (negative) "protective affect", and as a component in all psychopathologies.

Psychiatry in Sweden

As a medical speciality, psychiatry in Sweden has from the start and until our time, almost with no exceptions, had specially trained doctors responsible for the treatment. Others working within psychiatry – psychologists, mental welfare officers, nurses, keepers – have been subsumed under the term "paramedical staff". This implies that psychiatry is characterized by a hierarchical structure where the psychiatrist is at the top and, as already mentioned, has the ultimate responsibility for the treatment. Early on, therefore, the psychiatrist's view of treatment and of human beings decided the care the patients were offered. Before the great breakthrough of psychopharmacology – first neurolepticum (Hibernal 1950) for the treatment of psychoses – approaches to treatment in psychiatry may best be described as eclectic: the contribution of the psychiatrist included medical treatment; the task of the psychologist concerned diagnosis and possible psychotherapeutic treatments; the mental welfare officer considered social and family measures.

3 Alf Nilsson, 2005, 2009, 2011.

Psychiatry bloomed during the 1960s and the 1970s. Psychiatry open up for different types of treatments: social psychology, therapeutic communities, psychodynamic psychotherapy, psychotherapeutic divisional treatment of schizophrenic psychosis, theme treatment of initial psychoses in the home environment.

Since then a radical change has taken place within psychiatry. Nowadays, the medical-biological view dominates virtually the whole activity. This also implies a change in language, from using the expression "mental ill health" to "mental illness". I distinguish three factors that are probably important for this limited view: (i) The development of new methods for measuring functional patterns within the brain (PET, MRS, MEG etc.). This has resulted in an altered way of understanding psychopathology, from being states to becoming illnesses caused by changes in the brain and its transmitter system. (ii) The breakthrough of the American system of diagnosis *DSM* (*Diagnostic and Statistical Manual of Mental Disorders*). In the two first versions, *DSM-I* (1952) and *DSM-II* (1968), psychoanalytic and psychodynamic views still had some influence on the diagnostic categories included in these editions. With *DSM-III* (1980) the manual moved towards becoming totally atheoretical and descriptive. Each new edition – *DSM-IV* (1994), *DSM-IV-TR* (2000) and *DSM-5* (2013) – includes a substantial increase in the number of diagnostic categories and subcategories. In *DSM-5* there are around 350 categories. (iii) The complete breakthrough for psychopharmacological treatments for curing almost all psychopathological states. There are sources that show how psychopharmaca have become a tremendously large money-maker and have created a profitable market for the pharmaceutical industry.[4]

The SSRIs (Selective serotonin reuptake inhibitors) have a big impact. The claimed effect concerns the reuptake of the transmitter serotonin in the synapses of the brain. In spite of new generations of drugs there is still no evidence that they have any specific anti-depressive effect. In addition, the drugs function paradoxically by reinforcing and even creating the depressive state, and, so far, no SSRI has proven to work without side

4 Ingrid Carlberg, 2008/2011; Robert Whitaker, 2010/2014; Allen Frances, 2013/2014.

effects, which can manifest themselves differently in different individuals. To the latter it may be added that psychopathological states, in their whole variety, are connected with extremely complex transmitter substances and neurobiological patterns are involved, the latter patterns originate deepest from the brain stem's reticular system, and continue over the structures in the limbic system, the emotional brain, to a number of network formations in the cerebral cortex.

As is obvious in the epigraph at the start of this chapter, the psychiatrist Allen Frances is a sharp critic of the modern psychiatry and its one-sided biological and pharmacological development. I will clarify with some quotations:

> *DSM-IV* did not save normal, or protect it very well. Three years after its publication, drug company lobbyists won a huge victory over sensibler regulation. The United States become the only country in the world that allows direct-to-consumer advertising of pharmaceuticals. [...] No one dreamed that drug company advertising would explode three years after the publication of *DSM-IV* or that there would be the huge epidemics of ADHD, autism, and bipolar disorder – and therefore no one felt any urgency to prevent them.
>
> Evidence of diagnostic inflation is everywhere. There have been four explosive epidemics of mental disorder in the past fifteen years. Childhood bipolar disorder increased by a miraculous fortyfold; autism by a whopping twentyfold; attention deficit/hyperactivity has tripled; and adult bipolar disorder doubled.
>
> *DSM-5* has made it much easier to give and to get the diagnosis of Attention Deficit Hyperactivity Disorder. The experts [...] failed to consider that rates of Attention Deficit Disorder had already soared – so that now an astounding 20% of teenage boys get the diagnosis and 10% are on medication for it. In 1990, there were 600,000 kids on ADHD meds; now there are 3,500,000. Stimulant drugs that twenty years ago had annual revenues in the tens of millions are now approaching sales of ten billions a year. Positive proof that ADHD is way overdone comes from studies that the best predictor for having the diagnosis is your birthday – the youngest kid in the class is almost twice as likely to be tagged with ADHD as the eldest. Does it really make sense to turn immaturity into a mental disorder?[5]

5 Allen Frances, 2013/2014, pp. 73–74, p. 104, pp. 288–289.

As a contribution, my own commentary on diagnosing the behavioural disorder ADHD as supposedly brain damage. Once the diagnosis of ADHD is stuck, there is no other salvation than lifelong pharmaceutical treatment with narcotic-classified amphetamine pharmaceuticals (Metylfenidat, Concerta, Ritalin). It is the amygdala, "the jewel in the crown", in the limbic system that sets the affect tone from birth; the amygdala, which from birth is connected, via trajectories in the brain, to the hypothalamus, the seat of the autonomous nervous system, "the downwards connection" in a nutshell. "Upwards connection" exists in terms of the anterior cingulum and insula cortex, "the fifth lobe in the brain", necessary for face recognition, for instance. Moreover, there are trajectories, crucial for the regulation of affect, to the orbitofrontal cortex in the prefrontal cortex. It concerns two main trajectories, one for excitations, another one for inhibition. From birth, there is no visible "upwards connection" between the amygdala and the orbitofrontal cortex, "the convergence zone of affect regulation" (Antonio Damasio). It develops with its two main trajectories during the three first years of life, with two periods of growth spurts: 10–12 months and 16–18 months. So very different from the regulation of the (motoric) operations, from birth. It is ultimately associated with "the canonical neurons" in F5 in the premotor ventral cortex, in its turn with loops to the AIP (anterior intraparietal cortex). These are areas that, according to Giacomo Rizzolatti (see Chapter 1), provide the motoric system with a "repertoire" of actions that lay the foundation for later cognitive functions (operations) (see Jean Piaget in earlier chapters). Rizzolatti carefully underlines: (i) the regulation takes place without the involvement of affects; (ii) the loops between F5/AIP and the sensorimotor/subcortical areas in the brain are established from birth.

To conclude, a reminder of a debate within psychiatry in Sweden published in *Läkartidningen* (The Swedish Medical Journal), 2008, no. 105. It was initiated by two professors and chief physicians within psychiatry who claimed:

> that psychiatry is a speciality that should occupy itself with *illnesses* that are caused by *neurobiological changes in CNS, their diagnostics and treatment.* Psychiatry should

> thus occupy itself with *psychic illness* – not with different sorts of [...] interpersonal disharmonies. (italics mine)[6]

The article is commented by another chief physician and associate professor of psychiatry under the heading: "Psychiatry is a medical specialty – period!"[7] Would you, after a severe loss and in a state of pain verging on depressive grief, like to be received by a doctor and a psychiatrist who would be interested solely in your brain?

Personality and psychopathology

A brief historical résumé

It was Empedocles (483–424 BCE) who launched the idea about The All as being composed of four basic elements: earth, air, water and fire. Hippocrates (460–377 BCE) connected the four basic elements to the four types body fluids that he distinguished: air with blood, fire with yellow gall, water with mucus, and earth with black gall. Some of the persons that he took on to cure, Hippocrates thought, had too much blood, the *sanguines*, who we today would describe as being manic; others with an abundance of yellow gall, the *cholerics*, who were characterized by unregulated bursts of anger (personality disorder/psychopathy); yet others suffered from an abundance of mucus, the *phlegmatics*, a somewhat obsolete definition of such persons is neurasthenia (Freud); to the fourth category persons with too much black gall, the *melancholics*, were allocated (by disciples of Hippocrates after his death), in our terminology persons who suffer from depression.

The doctrine of the connection between body fluids and temperament has had a tendency to linger on in modern times. Academic psychology's

6 J. Dencker and J. Wålinder, 2008.
7 Göran Isacsson, 2008.

most prominent figure, Wilhelm Wundt (1832–1920), was prepared to embrace it. Also, the reflexologist and the founder of behaviourism Ivan Pavlov (1849–1936) concluded that the division into four elements was essentially correct.

According to Franz Gall (1758–1828), the standard-bearer of phrenology, it was the shape of the head and face that could be connected to the personality traits. In the wake of phrenology, doctrines about different kind of races flourished. It is hardly a century ago that research on "eugenics" was abandoned, most recently in Sweden, at Uppsala University.

Another opinion, that personality/psychopathology is connected to bodily constitution, has also lingered tenaciously on century after century. The German psychiatrist Ernst Kretschmer (1888–1964) created a doctrine about types based on differences in bodily types. Kretschmer distinguished between the slender (*leptosome*), the tubby (*pyknic)* and the muscular (*athletic*) bodily type. Kretschmer's conclusion was that a leptosome bodily constitution was the normal condition coupled with seclusion, delicateness, stinginess and a nervous disposition – in an extreme state, schizophrenia. The pyknic bodily state's normal condition was characterized by sociability, extraversion, agreeableness and intimacy – in an extreme state, manic depression. Furthermore, the athletic bodily constitution is characterized in the normal state by inertia, melancholy, contemplation and impetuosity – in the extreme state by an inclination for, on the one hand, depression, and, on the other hand, for violent outbursts, personality disorder/psychopathy.

Within modern psychological personality and psychopathological research, the approaches described above are today considered as historical parentheses.

Early on in the short history of academic psychology the opinion prevailed that it was the individual that was best equipped to express the "truth" about him- or herself, and his or her personality traits. The method preferred above all others was construed and more or less standardized formulas and self-assessment scales – that we even have not yet seen the end of today. The answers to the numerous questions and propositions that could be included in a form, then carefully underwent factor analyses. The result was more or fewer factors or dimensions that were regarded as revealing the person's temperament. Late in the 1990s a factor model was developed and tested on the human being's personality structure. It has resulted in the *Five-Factor-Model*, often referred to as *The Big Five*. The whole point of the model is that each individual's traits may be described with five, neither more nor less, factors or dimensions that vary in degree or intensity for different persons. The dimensions are: Neuroticism, Extraversion, Openness to Experience, Conscientiousness and Agreeableness.

Critics of this approach have pointed to various weaknesses. A factor analysis always gives results that depend on the data that are included in the analysis. Superficial data cannot result in anything other than surface factors. The critics have also focused on the fact that the narrative a person can convey about his or her "self" is limited. We may only get insights about ourselves to a certain limit. Beyond that limit there are layers that we never can reach through introspection.

It was Sigmund Freud who once and for all got us out of the illusion of total self-awareness. Consciousness, and particularly self-consciousness, is not master in its own house. There are layers in the individual's psychic equipment that are beyond the scope of self-observation.

I have published a whole book on the early Freud's topological construction of the psyche (Conscious/*Cs*, Preconscious/*Pcs*, Unconscious/*Ucs*) in relation to the levels of internal presentations that I distinguish in this book (*Srip*, *Seip*, System of Evocative Internal Images, *Seii*).[8] On the basis of a developmental perspective, extraordinary overlaps between these levels could be shown.

Freud was the first to formulate a clinically-based general theory on personality. The method that was connected to the theory was the psychoanalytic therapy with free association and interpretation of dreams. Therapeutic effect was possible through transference: the supplicant analysand/client always transforms in the relationship to the psychoanalyst/therapist elements that can be attributed to the early relating to the "primary objects" (Freud's denomination). Our life, when it comes to relationships, is never without a history. And what we once did not have the cognitive ability to reflect upon, we carry with us as adults. The domain (topos) where it

8 Alf Nilsson, 2009.

is gathered, and where it sometimes may live its own tumultuous life, is in *Ucs/Srip*. There we need to seek ourselves in order to catch a glimpse of why we in our adult life experience more fear, distress, dissmell ... than what the Good Life ought to offer us.

To seek backwards and scrutinize what it is all about on an unconscious level is often not a simple sinecure. Each topos defends its territory with all possible sorts of tricks and wiles that are at hand: warding off strategies of different kinds – because each topos, albeit deranged, should be able to keep (at the price of flexibility) its specific peace of mind.

After Freud, different personality psychologists and clinicians have borne in mind that a person is never transparent to him- or herself. In order to get some insights into how it can be in the layers lying beyond the visual field of a person's own experience, a large number of "projective" methods have been constructed – all with the aim of getting the person to reveal, indirectly, what dramas and traumas are there, in the deepest and most unlit of basement vaults. The most famous methods are probably the Rorschach or the ink stain test (originator Herman Rorschach, 1884–1922) and TAT (*Thematic Apperception Test* – originator Henry Murray, 1893–1988, Silvan Tomkins's mentor during his stay at Harvard University, USA). TAT is comprised of a number of pictures, each representing a key situation – most often with reference to human interrelations – that the person/client gets to watch. The task is to tell what he/she perceives in the picture. The narrative usually includes key experiences from the person's early life.

The characteristic of the Rorschach test and TAT is that the narrative presented by the person is always filtered through cognition (reflecting abstraction). This can contribute to an involuntary "intellectualization" (conditioned by what Freud called the warding off "rationalization" or "isolation") of the narrative. "Intellectualization" may overtake and overshadow what really might be going on in the person's deepest affective/recognitive emotional layers, forever inserted in and influenced by the contingency S-R-SE.

A brief account of psychoanalysis in a post-Freudian perspective

The cognitive tradition has always disapproved of Freud's psychodynamic thinking and psychoanalysis, and cognitive science still does, in spite of the fact that – or maybe just because – the thinking has had a large impact on the research conducted in many fields within cultural studies and the humanities in academies all over the world. Within psychoanalysis after Freud, one may distinguish different orientations: Early on in the USA, the so-called "ego psychology", and later "self-psychology"; in France the "semiotic" orientation with Jacques Lacan as the prominent figure; in England, the "object-relations school" with prominent figures like Melanie Klein and Wilfred Bion. Generally, this theorizing may be described as a kind of superstructure on the late Freud's speculative constructions "Thanatos/ the principle of death", Eros/the principle of life and/or his (clinical and atheoretical) "structural model" (Id, Ego, Superego). How this theorizing may express itself, I have previously criticized and dismissed as a possible way forward to a biopsychological human theory.[9]

Freud welcomed Charles Darwin's evolutionary thinking. To my mind, Freud never explored it fully and did not let it influence his psychoanalytic thinking. Instead, he took as his starting point the religious records and mythologies (the horde, the clan, totemism, the murder of the father etc.). In that sense, Freud was closer to the humanities than to the (natural) sciences. This applied for a time after he had given up his "neurobiological project".[10]

In a recently published edited volume, I've tried to trace the theoretical extensions beyond the psychoanalytic theory and its "self-sufficiency".[11] One way could be to connect to the "dynamic systemic approach" and/or the modern evolutionary thinking in terms of evo-devo as an overarching frame of reference (see Chapter 2). On a more limited level, a great deal of new knowledge has supervened in the field of the development of infants

9 Alf Nilsson, 2009, chapter 2.

10 Sigmund Freud, 1893–95/1994.

11 P. Fonagy, R. Krause and M. Leuzinger-Bohleber (eds), 2006/2009.

and small children.[12] In three of the seven authors in the volume, I find the tendency to incorporate approaches outside ordinary psychoanalytic theory. Marianne Leuzinger-Bohleber tries to integrate clinical psychoanalytic knowledge with *Embodied Cognitive Science* and its distinction between "declarative (autobiographical) and procedural (implicit) memory". The author refers to new neurobiological research on the hippocampus and its importance for human memory formation. Her conclusion is that, "'embodied' remembering is not simply 'non-verbal' or 'descriptively unconscious' but highly constructive, dynamic, and historically determined."[13] Susan Coates writes about *Childhood Gender Identity Disorder* (*CGID*) (DSM-IV). To understand this rare disorder, Coates refers to *imitation* of the absent parent-person's (the mother, in the case of the boy) "emergence and gestures". This reminds me of Teresa Bejarano's and the cognitive tradition's aetiology. Rainer Krause wants to see a shift from the (sex) drive to the affects as an explanatory ground. In this connection, he distinguishes three domains – attachment, autonomy and sexuality – in the development of the child that are important to consider. As the fundamental negative affects he discerns fear, anger, disgust and contempt. His point is that even if the affects are derivations of the drive, it gives a better clinical explanation if the drives and the affects are considered in the opposite order.[14] In all, I cannot discern any revolutionary contributions in the authors' otherwise traditional psychoanalytic theorizing.

Epigenesis: constraint as warding off

If we take Terrence Deacon at his word (see Chapter 2), order and organization are always connected with restrictions or *constraints*. For Deacon, then, constraint is never added to something else. Instead:

12 Alf Nilsson, 2005, 2009, 2011.

13 Marianne Leuzinger-Bohleber, 2006/2009, pp. 64–65.

14 Susan Coates, 2006/2009; Rainer Krause, 2006/2009.

> This concept of constraint can provide a negative approach to realism [in its accepted philosophical meaning], though it might also be paradoxically described as a *nominalism of absences*, since it is determined by discrete interaction effects that don't occur. (brackets mine)[15]

For you as a reader, what probably is paradoxical in this disconnected quotation is the last clause: "determined by discrete interaction effects that *don't* occur" (italics mine). Let us start by scrutinizing what "discrete interaction effects" stand for. To my mind, it is about interaction effects between components, or systems, that are delimited from each other through a *digital* "process". As has been made clear earlier, all delimitations of components and systems – thus, even the delimitations on the analogue line of affect/ internal presentation – that have taken place/take place during evolution are digitally created formations of interface. When I refer in the book to the "discrete" system formations, for instance, the bio-psychologically inborn *affect system*, as being separated from *emotions* in *Srip*, these two modes in turn being distinguished from *feelings* in the *System of Evocative Internal Images, Seii*, I mean the interfaces between them, which are some sorts of "ultimate varieties" (Deacon) in the shape of constraints. What I then denominate, considered in the phylogenetic perspective, "digital *processes* of interfaces that have *occurred*", it is something we know occurred, but we do not know *how* it occurred. These "spontaneous changes" that cannot be linked with any "validity behind", yes, these processes lay within the framework of what Charles Sanders Peirce called "tychism". There is thus also room for "interaction processes that don't occur", as Deacon would call "absentials", that we do not know anything about, but that might perhaps occur in the diachronous world where we live our lives on earth.

Let me take a purely hypothetical example from nature and phylogenesis to illustrate the implication of what I just claimed. The starting point is *Homo* who is without knowledge about frost-heat transformed into scales on a thermometer, but also ignorant about how to make a fire and use it as a source of heat (*Homo*, let us say, around two million years ago). The circumstances gave him the opportunity to perceive water as a liquid (to

15 Terrence Deacon, 2012, p. 191.

drink, as something in a watercourse, as rain); but probably also water in the shape of ice; but probably not in the shape of steam or water vapour. For *Homo*, on this level of development, there was an *inherent constraint* operating in the understanding that water, besides liquid and ice, may take the shape of steam. His "propagations of constraint" (Deacon) had not yet reached the "scale level" for the possibility of water to transform, via the formation of interface, into water vapour. Notably, water in nature counts as a digital system.

In physical nature – but also in all life – there are always "high level fluctuations" (Deacon), for instance, the waves in motion in the sea, that are different from "low level fluctuations", as for instance, the movement in a single water molecular. The first example can be observed with the naked eye (or by closing the eyes and listening to the sound in the form of wave lapping). For the latter example, specifically manufactured instruments are needed to be able to observe the condition. This may also be described in terms of "micro differences" (on the level of the molecular), that we cannot apprehend among the "macro differences" in our ordinary life which are within the framework of our apprehension. With reference to Peirce, Deacon sums it up in the following way: "The general properties by which we assess the similarities and differences among the diverse phenomena of our experience (e.g. pattern and order) are a consequence of the interaction between habits of our brains and habits of the world." To this we may add that "[habits] are [...] the ultimate locus of causality".[16]

But now we are dealing with "constraint" in relationship to "warding off". Constraints both delimit and tie together (hindrances that always are connected to relations, see above). As constraint, it is always "less variable than possible". As I interpret Deacon, it may be formulated in terms of a constraint always having a possibility of further variety or additional extensional scales (nowadays we may count in "scale steps" in both analogue and digital systems, that is, "accumulations"; compare with the "discrete mathematics" above). But in the life we live on earth, I am also inclined to interpret it as meaning that constraint, expressed in this way by Deacon,

16 Terrence Deacon, 2012, p. 202.

at the same time is connected to a sort of "optimal value in a statistical value scope", what we call "constancy" and even "invariance". Because constraint understood as "inherency", simultaneously lies behind "order" and "organization". Inserted into a developmental perspective, then, constraints contribute to the effect "that differences in one system can produce differences in the other, [...] and even in the remaining systems". This was expressed in a slogan-like formulation by the Nobel laureate in physics, Philip Anderson, as: "*More Is Different.*"[17]

To sum up, with reference to "constraints", which I associate with "the formation of interfaces", Deacon formulates a "negative approach" for emergence, thus something that is not added to something else (to add is a positive formulation of emergence). At the same time constraint/formation of interface are, to use the notion of the Nobel laureate Robert Laughlin, "protected states". As such states, they are elements that both delimit and tie together components and systems within a "statistical variation scope", with the best possible ability to make constraint/interface effective. This is valid for the systems in nature, as well as for organic systems, and for psychic systems, not least of all.[18] Is, then, constraint the same thing as warding off? Already in my first book, "The Mechanisms of Defence ...", I'm suggesting that warding off is an "extended" state of a normal delimitation state.

The phenomenon of warding off, in my first suggestion, is linked to constraints/formations of interfaces between the systems that successively during onto/epigenesis would emerge on the line of affect/internal presentation. Then it concerns transitions between emergences of four systems: (i) the affect system, (ii) the System of Recognitive Internal Presentation (*Srip*), (iii) the System of Evocative Internal Presentation (*Seip*), (iv) the System of Evocative Internal Images (analogously with the other acronyms it may denominated with *Seii*). If constraints/formations of interfaces between the four systems are important and even necessary for us in order to function "normally within a certain variation scope" within the human world, then, my second suggestion is that wardings off are inherent pathogenic "extensions" of constraints/formation of interfaces between the four

17 Terrence Deacon, 2012, pp. 198–200; Philip Anderson, 1972.

18 Terrence Deacon, 2012, p. 200; R. Laughlin and D. Pines, 2000.

systems mentioned above. As pathogenic they are "self-amplifying", inserted into diachronous time during epigenesis in order to unburden the child from an excess of negative affects. But as a formation in adult life they are "stiff" and "restrictive", and also "retained" so as to deal with (the negative) affect dynamic that appears, and ultimately manifest itself in psychopathological syndromes. But before I move on with constraints/interfaces as wardings off, I will go back to a suggestion that I made in my fifth book in Swedish about eventual differences between constraints/formation on interfaces as regards operative development on the one hand, and affect/internal presentational development on the other hand. My preliminary suggestion was to refer to constraints/formation of interfaces on the operative line in terms of "strong"; on the affect/internal presentational line in terms of "weak". In this book, by presenting two characteristics of the structural/functional brain as a background, I am inclined to preserve my earlier preliminary suggestion.

We already know that the operative development from birth with the inborn reflexes, nourished by expectation and the biphasic cycle, are connected to the cerebral cortex. In Chapter 3 I refer to at least two areas, F5 and AIP. How is it, then, with the innate affects? In earlier books in Swedish there are detailed descriptions of the trajectories, bottom-up, in the brain for the regulation of affects, all the way from affects at one end to their shape of feelings at the other. Two important and, by all indications, crucial areas for this regulation are the amygdala in the limbic system and the orbitofrontal cortex in the frontal higher-order cortices, which Antonio Damasio terms "the convergence zone of affect regulation".[19] From the beginning the newborn child lacks loops between the amygdala and the orbitofrontal cortex. They are developed, described in detail by Allan Schore, during the infant's first three years of life, in the form of a loop for excitation and other one for inhibition; with one growth spurt during the period 10–12 months and, another one during the period 16–18 months (see above).[20] To that, the knowledge about the existence of both "strong" and "weak" synapse formations in the brain has supervened.

19 Antonio Damasio, 1999/2000.
20 Allan Schore, 1994.

Against this backdrop it is not totally unreasonable to assume that the constraint/formation of interfaces on the operative developmental line is "stronger" than those on the affect/internal presentational line, which are "weaker". What we speak of in terms of "regression in the psychopathologies", I suggest, makes its way where constraints/formations of interfaces are "weaker", thus, "downwards" along the line of affects/internal presentations.[21]

To that I would only like to remind you of an issue that I discuss in Chapter 3 in this book: the line of operative development may be derived back in time to the evolution of the rhythmic activity/biphasic cycle with the emergence of the vertebrates (plenty of time for formation of the "strong" executive ability?); affects with the substantially later emergence of the mammals during evolution (less time for formation of the "weak" affects/relations?). Coming this far, I still insist: it was with the emergence and development of the affects that *Homo* became *Homo psychicus* – to the expense vulnerability and "weakness"? But once the "fusion" – "a loss of properties with ascent of scale" (Paul Humphrey) – of the operative system and the affect system is established, then there is in fact no return to a complete retrieval of the original elements in the form of, on the one hand, operations as such, and on the other hand, affects as such.[22] This "fusion" would then, according to what I suggest in the book, start at a "micro level" of constraints/interfaces where I have proposed that "surprise's contagion of expectation" probably takes place. Where I am when I write this, I think that we probably all do in our adulthood: both surprise and expectation are features that can manifest themselves within me – "whether it is in the form of feelings or thoughts, I do not care very much" – as a feature in my

21 Readers who have any knowledge in the field of physics will know that four forces are distinguished: electromagnetic force, gravity, and the "strong" and "weak" nuclear power. The "strong" nuclear power, you know then, is that with "action power", for example, in nuclear weapons; the "weak" nuclear power makes it possible for, say, quartz particles and other "microparticles" to "interact" and exchange energy, mass and charge with each other, that is to say, mutually interacting without the nuclear nucleus being affected. If nothing else – it can be a surprising analogy, admittedly expanded to breaking point.

22 Paul Humphrey, 1997, cited by Terrence Deacon, 2012, p. 199.

"me-ness". Largely, "I know" – as we generally do in our adulthood – what on the one hand is "thoughts", and what on the other hand is "feelings" in me. What is thinking and what is feeling, is, like your "me-ness", not always easy to tell, because our inner silent speech usually does not differentiate between thinking and feeling.

Perception and the Lund school with percept-genesis as an approach

A new philosophical view of visual perception as an introduction

Let us start this section with a modern philosophical view of visual perception as presented by John Searle in his new book *Seeing Things As They Are: A Theory of Perception.*[23] Searle calls his stance Direct Realism. This stance – and I agree with him mostly – I prefer to display by means of a number of quotations:

> Perceptual experiences have the mind-to-world direction of fit and the world-to-mind direction of causation.
>
> We start with the fact that the conscious perceptual experience has *presentational intentionality* built into it. And we can explain the actual content of the intentionality in terms of the *causal* component in the experience itself and its relation to the object in the world. (italics mine)
>
> Intentionality is that feature of the mind by which it is directed *at* or *about* or *of* objects and states of affairs in the world. Hunger, thirst, beliefs, perceptual experiences, intentions, desires, hopes, and fears are all intentional because they are about something.
>
> I think [that the special character of perceptual intentionality] derives from their special differences from other forms of intentionality. They are [...] along with

23 John Searle, 2015.

> intentions-in-action, hunger, thirst, and *raw emotions*, the most basic forms of intentionality. Most of their special features derive from the fact that they are *presentational, not just representational.* (italics mine)
>
> There are indeed shallow, unconscious mental processes, and these are often of serious importance, especially in questions of *human motivation.* (italics mine)[24]

John Searle is a philosopher. It would not be a particularly important question to him seek what is hidden behind "intentionality". He seems to be quite satisfied to study one, and only one factor causing perception to direct itself mind-to-world, namely "intentionality". But as a psychologist, as you as a reader have already figured out, this is a key question for me to try to answer. That the hunger drive could be one of the major elements behind an animal's "intentionality" would not be a surprise to you. And it would be quite clear to you that my answer in this book is somewhere in the clash between *affect* (*surprise-startle*) and *expectation.*

Homo psychicus and perception

In my first book in Swedish about *Homo psychicus* (2005), I derive and put perception in a general scientific context. In the second book (2009) there is a statement and description of the research on "subliminal perception". The gist of this is that subliminal or implicit perception is a scientific fact.

In my first book in Swedish I also argue for distinguishing *sensation* – which utilizes signals – from *perception* – which utilizes Gestalts. But I distinguish myself in particular from common scientific perception research by proposing that perception is *two-layered.*[25] As we know, perception involves the unconscious level of *recognitive internal presentations* and the "conscious" level of *evocative internal presentations* (Figure 6.1, drawn from Figure 3.1).

In addition to this "space proposal" I am now inclined to suggest a proposal of visual perception as related to *time*: the continuity of visual

24 John Searle, 2015, pp. 136, 132, 13, 60, 215.
25 Outlines most fully in Alf Nilsson, 2005, chapter 6.

perception might be based on *a running evocation of "synchronous/now/ Augenblick" that could be our basis of experiencing time as diachronous*. It does not exclude the condition of visual perception to be triggered of the synchronous/now/Augenblick in a situation whereas something *new* or *strange* appears. Thus, it is always preceded by the triggered surprise-startle.

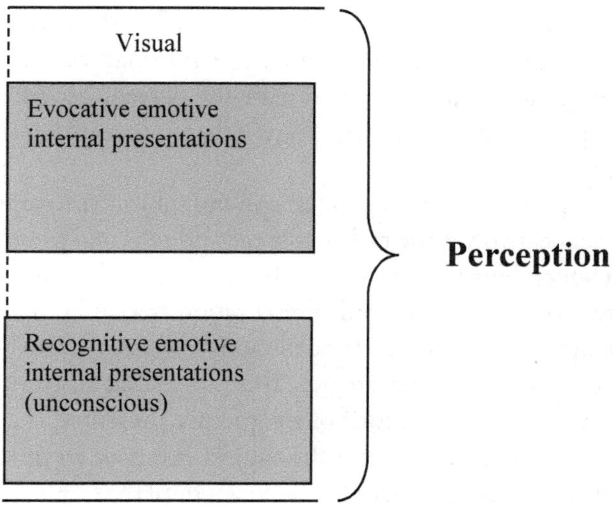

Figure 6.1. Illustration of perception as involving two strata.

Let us make it clear immediately why I above put "conscious" in quotation marks: In the dream, the level of evocative internal presentations has the status of being "preconscious". That the dream, in a perspective of perception, is of an evocative character is a precondition for us being able to recall the dream when awake. The recognitive internal presentations as primary "value gestalts" are, as was made clear above, never accessible as such and doomed to remain within the interfaces of *Srip* – and only indirectly accessible through their influence on the evocative internal presentations, images, feelings, thinking and/or language (language, for example, in dynamically directed psychotherapy).

Earlier I have described perception as an "immediate experience". The notion of "immediate" is twofold, expressing the "direct" nature and the temporal aspect of experience, on the one hand, and (as viewed literally) the "non-mediate" or "non-communicable" nature and spatial aspect of it, on the other. Seen in the *temporal* perspective, according to Silvan Tomkins, the affects function as the very cement for linking the present with the past and thereby are also highly important for the creation of the individual sense of coherence and same-ness over time (cf. above). Therefore, in the temporal and the developmental perspective, the immediate experience may be characterized as a domain where "fusion" of the present and the past as well as an organization of subjective internal presentation take place. It would not be wrong to designate the percept-genetic approach as being "experimental phenomenology" (thereby accepting the feature of affect as involved).

A technique used to capture the elusive and subjective nature of immediate experience involves the tachistoscopic and repeated presentation of pictures in which human interactions that are assumed to be emotionally charged are shown. Tachistoscopic presentation at a subliminal level prevents the subject's perceptual but unconscious *recognition* from being dealt with by operative cognitive strategies. This means that an eventual threat inherently linked with the motif of the picture presented is considered by the subject as actually present; the subject is unable to transform the induced threat into a *cognitively induced threat* until the C-phase level, that is, the level for correct perception, is reached. The presence of an "immediate threat" in the percept-genetic test situation also forces the subject to take prompt measures to cope with it. These measures are expressed as evoked or reproduced perceptual configurations/Gestalts – the subject's task being to verbalize and sketch what s/he *saw*, after exposure of the picture.

Visual perception has traits of both nowness/hereness and continuity. We have neither direct sense nor knowledge of how the latter is achieved. Above I describe it as being the concern of the affects/emotions. This is not a guarantee for experiencing a continuity in "me-ness". What we know is that preschool children have not yet achieved a sense of continuity with the "correctly stratified order" of the past (Alison Gopnik, Chapter 4). It indicates that the sense of "me-ness" probably has a "gap" within the preschool child. In this context, it may be of interest to shed some light on the preschool child's experience of gender identity. Infants between six and twelve months of age watch a picture for a longer time if it shows a child of the same sex than a child of the opposite sex. At the age of two, the child can distinguish between a dressed girl doll and a boy doll. If one shows the dolls without clothes, the two-year-old does not understand what separates the dolls. Six months later, when it comes to (dressed) adults and children, the child can verbally distinguish between "girl" and "boy". It is remarkable that only half of the five-year-olds understand the role of the sex organs in the separation of gender. It is only at the age of six to seven years that all of the children understand that the difference is anatomically based.[26]

To conclude, it is not until school age that the child achieves a relatively stable gender identity and time orientation regarding the past – where at least one elementary constant exists in "me-ness". I associate this with three points in the child's development that are about to take shape: on the operative line of cognitive operation, concrete operations; on the line of affect/internal presentation, the child's acquisition of primary evocative images, and of affects in the form of primary feelings. The child is in a transitional phase from "the stage of perception" as "superior", towards a stage when "operative cognition", "evocative image" and "evocative feeling" are on their way to becoming "superior".

Lev Vygotsky was early in having insights, on a general level, into the importance of the affects/emotions/feelings for the "phantasy life" (see above, Chapter 4, partly repeated here):

> Psychology calls this influence of the *emotions* on combinatory fantasy the law of general emotional sign. [...] A combined product of the imagination [evocative internal presentation/image] is generated, which is based on the common *feeling* or common emotional sign uniting these diverse elements that have become associated [...] *Affective* similarity unites and ties together objectively dissimilar images. (parentheses mine)[27]

Vygotsky has also, like few others, emphasized the importance of the sensory modality of vision during ontogenesis:

26 Susan Coates, 2006/2009, p. 104.
27 Lev Vygotsky, 1933/2004, pp. 18–19.

> The major direction of the child's evolution involves a steady increase of the role of vision in the enterprise of mastering the world. From its initial subordinate position it gradually becomes dominant and the child's behavioral motor-tactile system becomes subordinate to visual. During the transitional period we may note a struggle between two opposing principles of the child's behaviour, which ends in the complete victory of the visual mode of perceiving the world. [...] The dominant role in the perception of the world and in the creative representation of this perception is now played by visual aspects. The adolescent becomes more and more visual [...], experiencing [the world] mentally as a complex phenomenon and perceiving in this complexity not so much the variety and presence of things [...] as the relationships among them.[28]

As is obvious from the term "percept-genesis", the Lund school has tried to create methods or techniques in order to "cultivate perception in isolation" through subliminal exposure, and in this context to study the person's/the personality's "common emotional signs" recreated in a processual context.

Percept-genesis as viewed in the perspective of perception as involving two strata

As we have already seen, a key concept in the percept-genetic approach is that of "immediate experience". And as we also have seen, at the moment when the subject is asked to report the perceptual experience s/he has had, the experience already belongs to the past. The approach, developed by the late professors Ulf Kragh and Gudmund Smith as leaders, linked up with psychoanalytic thinking at the start in the 1950s and was based on two assumptions: (i) perception and personality are linked to each other, both with roots in the unconscious; (ii) early and unconscious experiences are conveyed as preconscious elements in processual studies of perception.[29]

Evan Thompson (see Chapter 2) describes perception in a way that resonates well with the percept-genetic view:

> Human perception is primarily directed toward how intentions [read affects/emotions/feelings] are expressed in other human beings rather than toward objects of

28 Lev Vygotsky, 1933/2004, pp. 81–82.
29 Ulf Kragh, 1955; U. Kragh and G. Smith (eds), 1970.

> nature and their sensory qualities. Human perception grasps intentions as "experienced realities", not as objects in a detached or intellectual sense. [...] Perception is thus first and foremost physiognomic, and later it is directed toward things as use-objects. Perception never loses its physiognomic quality; it remains physiognomic not simply in its nascent phases, but in its mature phenomenal character ... (brackets mine)[30]

In visual perception, thus, the vantage point is always in the "physiognomy", enclosing the "psychic". Can this be comprehended against the backdrop of my two-strata theory of perception? In my fourth book in Swedish I have argued that "simultaneity" for the subject is a compound phenomenon of "compressed recognitive synchronous/now/in Augenblick" – plain and simple nowness/hereness. As Nicholas Humphrey underlines, newness/hereness includes a "circle of causation" with "an overall blending of cause and effect". He speaks of a "long-lasting sequence with the 'extended present'". The time for "the circle of causation" takes up to half a second, the same timespan that Benjamin Libet indicates that it takes from impulse to action.[31]

When Kragh and Smith at the beginning of the 1950s started to use a tachistoscopic technique for rapid exposures of specific selections of picture themes, the technique was not completely untested. It had been used by Gestalt psychologists and organismic oriented researchers – but also by Jean Piaget in the 1920s. My description is hardly wrong when I refer to Kragh and Smith as the first psychological researchers that used the technique in a more systematic way.

A brief interlude on Sigmund Freud's conception of "warding off"

Let us begin with a clarification of Sigmund Freud's conception of *Abwehrung*, "warding off", on the one hand, and *Verdrängung*, "repression", on the other. In his earlier works, he used the term *Abwehrung* as a general designation – later on, the concept of *Verdrängung*. In *Hemmung,*

30 Evan Thompson, 2007, p. 77.

31 Nicholas Humphrey, 1992/1999, pp. 85, 187; Benjamin Libet, 1992.

Symptom und Angst (1926) he argues in favour of reverting to the old concept of *Abwehrung*/"warding off" as "a general designation for all the techniques of defence the ego makes use of ...".[32] According to Freud, the psychotic patient utilizes the warding off of *Verwerfung/Verleugnung*, for which I have used the concept of "abandonment" in English.[33] In all kinds of "neuroses" Freud distinguishes an element of "repression", sometimes also named "suppression", as one "mechanism" among others ("denial", "reaction formation", "isolation", etc.). Later in his life Freud became interested in what we today define as "earlier disturbances/disorders". In a posthumously published paper, "Die Ichspaltung im Abwehrvorgang", he described the warding off "mechanism" of *Ichspaltung* that has been designated in English by the concept of "splitting" (see above Chapter 3).[34] Regarding my interpretation of Freud, the "mechanism" of "isolation" is the most advanced one he distinguishes as a warding off element in "neurosis". This is the last one intrapsychically for the patient suffering from the complex obsessive-compulsive disturbances, for internal warding off, when both repression/suppression and reaction formation have failed in their attempts to impede the intrapsychic process from displaying the basic impulse expressed as anxiety. For the obsessive-compulsive patient it is thereby an issue of anxiety in the guise of "guilt-anxiety" (Freud), in my terms "feeling of anxiety".

Viewed in the perspective of ontogenesis, abandonment applies to the amental infant until the first elementary formations of recognitive emotive internal presentations (about eight months) are established; splitting applies once the first recognitive emotive internal presentations have been formed, and from then on during ontogenesis; repression/suppression applies at the time when, and after, the visual evocative internal presentations have emerged; *isolation* from the stage in ontogenesis when operative cognition, *thinking*, is possible to distinguish from *feelings* (of anxiety), fully when the child has developed the ability for reflecting abstraction, simultaneously with her/his accessibility to the "feeling of me-ness", and thereby also with complete accessibility to "anxiety as feeling" (about eleven to twelve years

32 Sigmund Freud, 1926/2001, p. 163.
33 Alf Nilsson, 1983/1994.
34 Sigmund Freud, 1940/2003.

of age). The next step is to present, in a schematic arrangement, the four constraints/interfaces extended to wardings off (already *per se* presented in my early book about the "mechanisms of defence").

Constraint/interface → Wardings off	*System level*
Abandonment	Amental
Splitting	*Srip*
Repression/Suppression	*Seip*
Isolation	*Seii*

Do we know anything about the process on both sides of and in the transitions between the systems, where the different constraints/interfaces/wardings off in the scheme above are established? It is worth reminding readers in this context that even regarding the line of affect/internal presentations, the interfaces between the systems are digitally formed. Given that, with the knowledge we have today we cannot define the process; perhaps it is even completely undeterminable. If we follow Terrence Deacon, what is valid for those interfaces is: "ambiguous spatiotemporal origins, [...] that are only statistical and dynamically definable, and are defined by a dynamical quality, [...] and not a discrete extensional boundary."[35] Deacon's description is analogous to how R. Crease and A. Scharff Goldhaber define the interfaces at the quantum level: "Today, the boundary transition is known to be sometimes abrupt and sometimes gradual, involving a proliferation of many different happenings."[36] Despite these discouraging descriptions concerning possible measurability, I persist in the proposal that "the boundary transition" between the four system levels conceptualized in the scheme are mutually different at the various levels, and that there are different elasticities/transparencies between the pole where constraint ought to be a more correct term and the other pole, where wardings off is a more correct term. The foundation for this is that the operationalization takes

35 Terrence Deacon, 2012, p. 167.
36 R. Crease and A. Scharff Goldhaber, 2014, p. 220.

place in perception as defined above, utilizing on the one hand the level of recognitive internal presentation in its dynamically connection, and the level of evocative internal presentation on the other.

PORT: a projective method for clinical description

The technique of PORT uses an apparatus that consists of a regulator for time exposures and a viewer into which the subject gazes through a tube with a close-fitting eyepiece. The projected picture is 42 × 63 mm in size, the distance from the subject's eyes to the picture being about 30 cm. The exposure time is varied from 10 to 2,000 ms in a geometric series of twenty steps.

The pictures are exposed once at each exposure time beginning at 10 ms. Subjects are asked to report everything they see and to make a simple drawing after each exposure. The experimenter writes down the subject's verbal report. The test series is discontinued once the subject gives the correct report for three consecutive exposures, the first of which is scored as "C-phase", in normal groups on average at exposures 15–17. In the absence of a correct report, the test series proceeds up to the twentieth and final exposure ("no C-phase"). The three pictures or themes involved in PORT are described below. For a complete description of scoring categories, reliability and validity of the PORT, readers are referred to the Manual.[37]

Attachment theme. Influenced as it is by John Bowlby and modern infant research, the PORT theme corresponding to the child's first relationships to the mother persons is called the attachment theme. The drawing portrays a young woman (mother), wearing a dress but without shoes, sitting on a chair in a room. An infant, dressed in children's clothing, is sitting on the woman's knee. The woman's right arm supports the infant's back.

37 A. Nilsson and B. Svensson, 1999.

Separation theme. This theme is named after Margaret Mahler's description (cf. above) of what she referred to as the "second object-relation phase". It shows a young woman (mother) standing in a doorway to the right in the picture with her back to the room. The woman, dressed in a coat and holding a handbag in her left hand, stands with her head turned to an approximately one-year-old infant lying on the floor to the left with his or her right arm stretched out towards the woman. The infant is dressed in children's clothing. There are some toy bricks on the floor close to the child.

Oedipal theme. The picture here shows a young male (father), dressed in trousers and a polo shirt, standing at the left with his arms stretched towards a three- to four-year-old child who is standing in front of him. The child is dressed in children's clothing and turned with his or her back towards the male. There is a ball lying on the floor somewhat further away from the child.

Each PORT theme involves a number of major scoring categories. These categories, in turn, contain subcategories designated by the letters (a), (b), etc. An (a)-category is conceived of as being more severe and developmentally primitive than a (b)-category, etc. For example, anxiety, which is scored in all three themes contains three categories: (a) "complete dissolution anxiety, (b) "partial dissolution anxiety", (c) "blackening". The three categories or "signs" are scored: (a) when dissolution of the perceptual configuration previously reported in an exposure is completely dissolved in the following one – "only black and white dots and lines"; (b) dissolution of the configuration is partial, for example, involving only the child; (c) stimulus-inadequate dark or black formations appear in the reports of the picture – "feeling of anxiety". In the following named categories are presented in addition to illustrations of different reports for patients with specific psychopathological states (and diagnosis).

In the Manual the clinical materials investigated – in all, nearly 500 patients with different diagnoses, right up until the Manual was published (1999) – are presented under the following headings:

Initial studies

Psychogenic vertigo

"Neurotic disorders"

Psychosomatic disorders
- Oral galvanism
- Ulcerative colitis
- Bronchial asthma
- Rheumatoid arthritis
- Psoriasis
- Migraine

Substance-related disorders
- Drug abusers
- Benzodiazepine abusers

Schizophrenic disorders
Borderline disorders
Antisocial disorders
- Court-ordered treatment patients
- Patients with sexual abuse

Conduct disorders
- Five groups of juvenile delinquents

Eating disorders
- Anorexia nervosa
- Bulimia nervosa

Affective disorders
- Mood disorders
- Panic attack

In addition, samples of normal controls have been tested: 581 on the attachment theme, 1,018 on the separation theme and 177 on the Oedipal theme.

In the client's nowness/hereness we try to recreate his/her thenness/thereness: in PORT, first, with reference to his or her inner domain regarding attachment relationships; then, with reference to the experiences of separation/individuation/autonomy; lastly, with reference to the domain of early "Electra/Oedipal"/gender relations. The oral report and the drawing are, to quote Evan Thompson:

> a representation that abstracts and combines the contents of many attentional phases of visual experience. It is a static representation of a temporally extended, dynamic process of [...] mental exploration of the scene.[38]

Psychopathologies systematically follow epigenesis: Illustrations by PORT

I will begin by repeating the thesis that psychopathologies can be reliably described in an epigenetic perspective with reference in the line of affect/internal presentation and inserted into the contingency S-R-SE. I have earlier stated, as an overall thesis, that an excess of negative affective experiences at the expense of positive ones during epigenesis increases the prospect of psychopathological problems in adulthood; while the reverse, an excess of positive affective experiences at the expense of negative ones reduces the prospect. This thesis I shall now expand with the proposal: Overload of negative affective triggering during epigenesis is traumatizing and is something the individual within him- or herself strives to protect him- or herself against with the aid of constraints/formations of interface, in their extreme extended into forms as: abandonment/splitting/suppression/isolation as major wardings off – both simultaneously and/or *ad hoc*.

There are indications that the amental infant lacks clear boundaries between inner and outer. Following the cognitive theorist Alison Gopnik, for the infant the outer world is equivalent to one that the parents make the infant pay *attention* to. According to Gopnik, the child's internal world is a matter for the parents.[39] What importance the affects have in this context is not an issue that Gopnik deals with in her book. In the best of worlds, the gradient affect *interest* is, coupled with *enjoyment-joy*, the child's inner companion; in the worst of worlds, the traumatizing toxic and malignant gradient affect *fear* (and/or *anger-rage* and/or *dissmell* and/or *disgust*). If, during infancy, in the attachment-phase, an excess of fear-terror prevails at the expense of interest, an escape from abandonment is what the child then has at his or her disposal. Is this possible here, to rewrite it in terms of

38 Evan Thompson, 2007, pp. 281–282.

39 Alison Gopnik, 2009/2010.

onefold/the analogue and twofold/the digital? Note Wilden's proposal that interfaces within analogue systems always occur via digitalization. In the example, it is an issue of amplification of the stimulus and response in the form of the affect fear-terror, which at "its extreme border" – "the border of affect-escalation" – works as a code for digitalization of abandonment that would be a constraint in the formation of interfaces.

Schizophrenic psychosis

The illustrations that follows are from a study in which patients with a diagnosis of schizophrenia and patients with a diagnosis of borderline were compared using PORT.[40]

Figure 6.2 reveals an excerpt from the attachment series for a patient with the diagnosis of "catatonic schizophrenia" (ICD-9-CM). The designation in DSM is *catatonic behavior*. It is a deep state of schizophrenia, which is characterized, for instance, by a blocking stupor (catalepsy) with motor disability and physical apathy, alternating with motoric hyperactivity "without intended meaning" (monotony), but which may affect the environment in the form of violent attacks (on things as well as on people). This condition also involves an altered state of consciousness, in which the patient is difficult to get in contact with. Sometimes the patient, without intended communicative meaning, can repeat phrases and sentences (echolalia). The patient, a forty-eight-year-old man, was treated at a forensic clinic (after previously having been cared for in an open psychiatric clinic), because his condition was considered to be chronic.

From his report of exposure 1 this patient appeared to dwell in a cosmic sphere, which calls to mind Harry Stack Sullivan's statement that for the catatonic "the struggle expands to cosmic dramas", and Silvano Arieti's view that the catatonic's "feelings are reminiscent of the feelings of cosmic power [...] [and that such patients] believe that by acting they may cause the destruction of the universe".[41] In exposure 2 the same patient reported

40 Alf Nilsson, 1995.
41 Harry Stack Sullivan, 1962, p. 113; Silvano Arieti, 1974, p. 161.

"a room" and in exposure 3 "something demolished, the room, the house". In exposure 6 the patient for the first time reported "Empty space", which he adhered to until the last exposure (exposure 20). In the PORT manual the sequence is scored as "Regressive fragmentation".

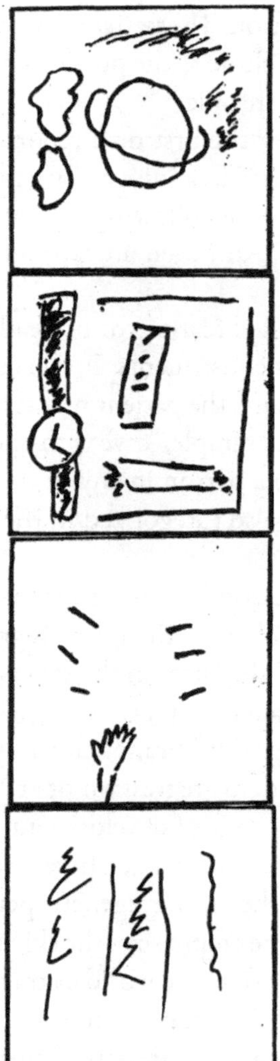

Exp. 1 'A planet with a ring around. A lilac sky.'

Exp. 5 'A picture of a wrist-watch. Room with a carpet. A black bracelet of the watch.'

Exp. 7 'Empty space. An outstretched hand.'

Exp. 13 'The same empty space.'

Figure 6.2. Example of the score "Regressive fragmentation" of "space" from the patient with a diagnosis of catatonic schizophrenia. Excerpt from the attachment series.

In Figure 6.3 an excerpt from the attachment series for a patient with the diagnosis of "hebephrenic schizophrenia" (ICD-9-CM) is shown. In DSM the classification is *schizophrenia of disorganized type*. This is also a deeply schizophrenic psychotic condition, which is characterized by long-standing withdrawal, disorganized speech and behaviour, as well as delusions about body dissolution. The patient, a man aged forty years, was treated in the same forensic clinic as the previous patient, because his condition was considered to be chronic.

In exposure 5 the patient for the first time reported a person in the picture – the person is exposed and in a "distress situation". In exposure 6 it would be "the same person". In the following exposure it is only a fragment of a person, "white arms". And then, in exposure 8 the report is a skeleton lying on a coffin.

It is probably about the naked fear-terror of death, where the body is destroyed and dissolved in its constituents. In the continuation of the series, up to the twentieth exposure, the patient reports different contents in each of the new exposures (for example, "cave with opening" in exposure 13). No one reports having seen a person in any of the exposures. In the PORT Manual, this content is also categorized in the attachment series as "regressive fragmentation".

The term "regression" is the opposite of "progression". In psychology (and in everyday life) one usually uses the term "regression" without a systematic link to "progression", the latter in the sense of "development": evolutionarily, ontogenetically and/or epigenetically. That the main perspective in this book is "development" cannot, at this point, has escaped the reader. Therefore, when I discuss regression or "remission", now and henceforth, it refers to a certain stage of development. Is it then possible to speak in terms of a specific stage for what I here refer to as "regressive fragmentation"? Considered in the onto/epigenetic perspective, I refer to in Chapter 3 to Didier Anzieu. He emphasizes the skin as being the child's first membrane with, on the one side, the world as fragments, and on the other side, the body as fragments. What may apply to the child during the attachment phase, I speak in terms of (affective) "onefold", according to Anzieu "cenesthesia", defined as: "a general sensation of existence in the world without boundaries between outer and inner". In Chapter 3, I also

Exp. 5

'It was somebody shouting something with his arms upwards. Perhaps a distress situation.'

Exp. 6

'It was the same picture.'

Exp. 7

'I didn't see what it was. Saw only white arms put out.'

Exp. 8

'I didn't see.
It's somebody dead.
Skeleton in some way.
It's difficult to draw a
skeleton but he was
lying on something,
a coffin or something.'

Figure 6.3. Example of the score "Regressive fragmentation" of "body" from the patient with a diagnosis of hebephrenic schizophrenia. Excerpt from the attachment series.

refer to Allan Schore. During the attachment phase his description is that "the infant's right hemisphere is 'psychobiologically attuned' to the output of the mother's right hemisphere". Furthermore, besides the fact that in these psychologically attuned meetings, eye-to-eye, the foundation for the child's affective regulation is laid, but also the foundation for the child's "body and space orientation" – both also ultimately regulated by the right hemisphere. Viewed in the epigenetic perspective, my interpretation is: for the catatonic schizophrenic patient there are deep flaws in an early attachment relationship without the establishment of a proper hold in the mothering world/the outer world; for the hebephrenic schizophrenic patient, deep flaws, without the body achieving a sense of cohesion that we usually experience our body to be as an adult. Modern cognitive researchers would probably talk about impairments in the patient's "*embodiment* process".

Figure 6.4 shows two exposures, one from each of two patients, and from their separation series. Their diagnoses were "paranoid schizophrenia" (ICD-9-CM) (at the time of the execution of the study, recommended by the National Board of Health and Welfare, although the diagnosis as such is not found in DSM). The two patients admitted to a psychiatric clinic are men aged twenty-nine and forty respectively. As a schizophrenia diagnosis, "paranoid schizophrenia" can be regarded as "in depth between catatonia/hebephrenia and 'schizophrenia unspecified', NUD (Non Ultra Descriptus)", the latter diagnosis also recommended by the National Board of Health and Welfare. I will return to patients with NUD diagnosis in the study.

The drawing in the first square (Figure 6.4) was reported by the patient: "Something like a chain" (I myself associate it with "a piece of barbwire" or "a suture/scar formation"). The second patient reported in the same exposure (6) "A chain I thought I saw".

Exp. 6

'The floor of some kind.
Something like that
hanging there.
Something like a chain.'

Exp. 6

'I chain I thought I saw.'

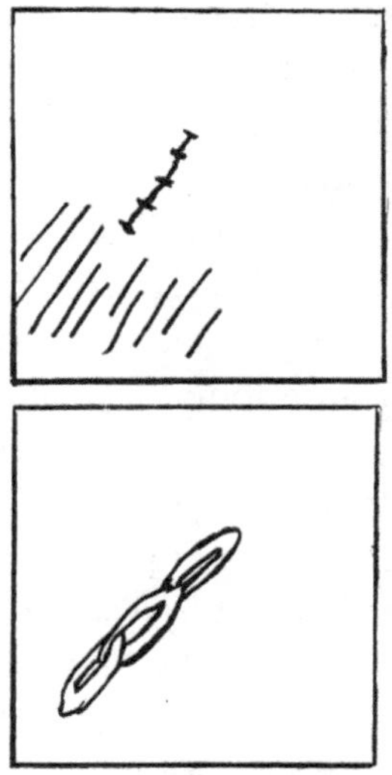

Figure 6.4. Example of "primary dependency as warding off". Two patients with a diagnosis of "paranoid schizophrenia". The separation theme.

In both cases, scored as "primary dependency as warding off", the chain appeared between the child and the woman even before human beings were reported in the picture – the relationship appeared "as such" before the patients reported persons in the picture. An immediate interpretation of the answers is that patients "evoke a strong imagination of bonding – where in reality only (internal) rudiments of bonding exist". At the same time, the fantasized bonding appears to be objectified and dehumanized – chain – which is of a deeply archaic kind. My impression is that the patients give expression to illusionary worlds that normally appear in dreams – but hardly ever occur in our ordinary world of imagination.

Against the backdrop of Figures 6.2 and 6.3, is it possible to tell what kind of intellectual and imaginative worlds these deeply schizophrenic patients are dealing with? For the catatonic schizophrenic patient, the illustrations correspond well with the outside world as being dissolved and impossible to get hold of (the catatonic state as the complete withdrawal from an intangible world, exposure 7). For the hebephrenic schizophrenia patient delusions are involved, about the body as being subject to ongoing dissolution. As mentioned above, the reported sequences here are the only ones in PORT that have a content that may be interpreted. For the patients illustrated in Figure 6.4, the PORT series are somewhat more cohesive. The content illustrated is retrieved from each patient's separation series. Epigenetically, it requires access to a kind of experience of separation/individualization/autonomy – albeit rudimentary; that is, access to contents on a recognitive emotive internal presentational level, in *Srip*, that show that relationships with caregivers are possible. However, for the patients illustrated in Figure 6.4, the relationships become somewhat dehumanized and "physically detached", without the support of any (caring) person (in the picture). Can we understand the contents of the schizophrenic patients' PORT series, viewed in the perspective of the general frame of reference presented in this book?

In the *worst* of worlds, where an extremely fragile attachment has been granted the individual, I trace features of abandonment as warding off in the aetiology of schizophrenic psychosis. It ought to have been a disadvantage for the approach between the affect surprise-startle and the operative expectation, by an early "gap" between the two lines of development in the continued epigenesis. Instead of interest-excitement as a gradient affect, an excess of fear-terror and anger-rage would have ruled. In the continued epigenesis, it probably counteracted the approach between surprise and expectation, later on the formation of recognitive emotive scripts by interest as admittance and enjoyment as relief in the "scenic" script formation in *Srip*. In the catatonic schizophrenic patient, it ought to have led to flaws in the attachment with a "projective" course directed towards the mothering world transformed into the outer world with physical things/manufactured objects/artefacts – in the regressive state with thought disturbances and as a consequence, internal presentations of the outside world as a cosmos doomed to disruption and fragmentation. In such a world, explosive anger as a *level affect* is the only thing that can counteract and defy the disruption and fragmentation. In the hebephrenic schizophrenic patient, the lack of attachment relationships with "introjective" extension towards the inner world – the body as "the whole world" – has led to disturbances of thought resulting in beliefs about the body as being non-cohesive. It causes the hebephrenic schizophrenic patient to live in constant fear of body fragmentation.

In PORT, the differences between the schizophrenic states become clear at different "levels". For the patients with a diagnosis of paranoid schizophrenia, as already mentioned, the series are more coherent and with contents that can be linked to the different PORT themes. These themes, in the study referred to above, are designed to seek the origins of the schizophrenic patient's delusions, which are of different kinds and degrees. "The primary dependence as warding off" in the separation series suggests that for patients, a serious dilemma about separation/individualization/autonomy is involved. It may be "an inner focus", a "stage" later on in affect and relationship development than that of the patients with catatonic/hebephrenic schizophrenia. I interpret this as meaning that warding off of abandonment did not bring about the same deep gap between the line of affect/internal presentation and the line of operations in patients diagnosed with paranoid schizophrenia. Regression in these patients manifests itself most clearly in the sphere of recognitive internal presentations. However, serious delusions are involved with elements of the level affect distress-anguish – which in the schizophrenic paranoid patient, through projection as warding off, may quickly slip over into anger-rage.

As a reader, I can imagine that you wonder how it is possible for the patients to *evoke* these delusions in PORT. In my response to that, I refer to evolution. There is no adult person without access to recognitive, evocative and cognitively operative abilities – even if they are insufficiently nourished during the course of epigenesis. The suggestion is that they exist as a predisposition in the embryo during the foetal stage – the human embryo that always leads to the birth of a human being. If these predispositions were missing in the embryo, the creature being born would never reach the stage of an adult human being.

Borderline, narcissistic, antisocial, etc. personality disorder

The designations of the psychopathologies in the heading are retrieved from DSM. If I initially limit myself to "Borderline", it is clinically a problematic term: "Not exactly this, psychosis, and not exactly that, neurosis." Although the "discovery" of this condition is usually attributed to psychoanalysis, it was early adopted by descriptive psychiatry. In DSM it is included under the heading "Personality disorder, type Axis II". For the diagnosis to be valid, at least five out of the eight criteria must be met. The psychoanalyst and psychiatrist Otto Kernberg is perhaps the most renowned among those who have made advanced descriptions of "early disturbances"; borderline he termed "borderline personality organization (BPO)". Kernberg inserts BPO into a "conflict model", while the primary representative of "self psychology", the psychoanalyst Heinz Kohut, places the disorder in a "lack model". If we follow Kernberg, this concerns patients/persons with: (i) deficiencies in identity integration, (ii) remaining ability for reality testing and (iii) splitting as a basic warding off strategy.[42]

With the purpose of confirming the clinical value of PORT, diagnosed groups have been used in the studies, often with reference to two or more diagnostic models. The results have usually confirmed (validated) the criteria set up for the models. The results have at the same time demonstrated that the acknowledged criteria for diagnosing psychopathologies appear to be inflexible and sometimes even rigidly square; patients are unique individuals, each with both general and specific character profiles in PORT.

In Figure 6.5 there are two excerpts from two of the *borderline* patients involved in the study of patients with schizophrenia and borderline. Both patients report something so deviant from the displayed picture content as – a face! "Face" in the attachment series is scored, according to the manual, as "whole configuration as warding off". Nine out of twelve borderline patients in the study report perceiving a face, while another two of the patients are coded for "whole configuration": one reports seeing an ear (Figure 6.6); the other reports seeing a "double bass". How should we understand the reports

42 Otto Kernberg, 1975, 1976, 1984; Heinz Kohut, 1971, 1977.

of "face" within the frames of the attachment theme? As I have repeatedly said in this book, the face is a basic configuration – the first of interest in the world for the newborn child and the first visually evoked imagination of the three-year-old in the act of depicting (the cephalopod). As already mentioned, the face is special to Man. In PORT "face" is – like other whole configurations – a different and deviant response among normal individuals (6 per cent in a group of 581 non-clinical individuals compared to 11 out of 12 in the borderline group). In addition, the borderline patients in the attachment theme showed a *sequence* of a sign of whole configuration in one exposure followed immediately by dissolution anxiety ("anxiety a"). This was the case for ten of the patients (none of the 581 individuals in the non-clinical group). If this concerns the fear of body dissolution in the schizophrenic psychotic patient, it concerns fear of *psychical dissolution* in the borderline patient.

Exp. 10

'A little more face now.
Nose, mouth, half-long hair, dark.
More sharpness.'

Exp. 11

'It was a man who pulled faces.'

Figure 6.5. Examples of face as "whole configuration as warding off" for two borderline patients. The attachment theme.

In Figure 6.6 examples of whole configurations other than faces appear in PORT's attachment theme. In the first case, it concerns a man with severe and prolonged alcoholism, at the time of the test, taken in urgently for detoxification. The second example concerns a woman, also having problems with drug addiction and taken in for treatment.

Exp. 7 'Caramel custard or ice cream in a dish. A strong gleam from it.'

Exp. 13 'Dusky. The whole picture is transparant. An ear as background figure.'

Figure 6.6. Examples of "whole configuration as warding off" from a male alcoholic and a female drug abuser, respectively. The attachment theme.

If it is an issue of borderline distortion/pathology/organization, there should be signs of splitting as warding off. I have previously described it as meaning that in *Srip* a *delimitation* occurs of "categories for distinguishing value patterns" that (more or less) resemble each other – of faces, more or less specific and strange (resembling familiar faces).

As appears above, categorization in the analogue is always about digitalization for formation of interfaces to appear. Now my suggestion is that "normal" digitalization that results in "delimitation" has an important adaptive function. Precocious pushing forward of "delimitation due to charge of negative affects" in epigenesist makes the child always be at the forefront of the next opportunity in development. It may cause the delimitation to adopt the warding off function of splitting. As such, it is a structure that is simultaneously both fragile and rigid. My proposal includes a slide towards the interface of the next imaginative opportunity in ontogenesis, that is, the System of Evocative Internal Presentation, *Seip*. Thus, if in the epigenesis it is about precocious development of delimitation → splitting, it should also apply to the shift towards stereotype splitting. Instead of *Srip* as an open, flexible and "unprejudiced" – Ignazio Matte Blanco – value system, it should lead to a more rigidly polarized system of either "completely out of good", "idealization" or "completely out of bad", "devaluation" "– borrowing" the language of psychoanalysis and Matte Blanco.[43]

In a study of LVU juveniles (LVU is the abbreviation of the Swedish name of the Care of Young Persons Act) it was the first time splitting was scored in PORT. At a one-year follow-up, it was shown that the juveniles who had scored splitting had more serious features of abuse and crime than other juveniles in the group.[44]

In Figure 6.7 there is an illustration of the two variants of splitting (a) that were scored in PORT. The illustration is selected because both variants are consecutive in one and the same separation series. As shown in the figure, the drawing is *sketched* with a vertical line, or the door is moved down to the bottom of the picture, so that the woman and child end up in two separate fields.

Of the twelve borderline patients in the study, ten were scored for splitting. Moreover, the two other patients in the borderline group were the only ones who showed signs of severe depression (a), see below. Among the schizophrenic patients in the study, three were scored for splitting in the separation theme. It is remarkable that it was the three patients that had been diagnosed with NUD, literally the three patients "in the borderland" between the diagnoses schizophrenia and borderline.

43 Ignacio Matte Blanco, 1975/1998.
44 A. Nilsson and B. Svensson, 1984/1999.

Exp. 14

'Someone creeping.
And a woman too.'

Exp. 15

'A woman goes
through a door.
And a baby creeping.'

Figure 6.7. Illustration of "splitting" from the same series in PORT. The separation theme and the child end up in two separate domains in the drawing.

A third sign in PORT, predominantly concerning the borderline patients, is also coded in the separation theme, and is termed "transitional object". A common answer is "a teddy bear". Other responses coded are "a security blanket", or "a doll", or "a rag doll", or "a feeding bottle" etc. Nowadays, transitional objects are probably widely known. As appears above, the paediatrician and psychoanalyst Donald Winnicott coined the concept (see above, Chapter 4). In a developmental perspective, Winnicott argues that access to a transitional object is important for the later separation between *me* and *not-me*. In the psychopathological perspective, with Winnicott's language, it may be about "excessive fixation" as a substitute and/or comfort with reference in deficiency during early epigenesis. An example of scored transitional object is shown in Figure 6.8.

The study of schizophrenic and borderline patients revealed that nine patients in the latter group and three in the former reported a transitional object in the separation series. Among the schizophrenics, two of the patients were diagnosed with NUD, the third with paranoid schizophrenia.

Exp. 20

'A teddy bear on the floor.
The woman is standing
in the doorway.
The child on the floor is trying
to crawl forwards.'

Figure 6.8. Illustration of a "transitional object". The separation theme.

During a visit for research purposes to a clinic for patients sentenced to psychiatric inpatient care, a large number of patients were tested with PORT. Patients were grouped on the basis of crimes and diagnoses. Ten patients with a diagnosis of *psychopathy* were compared to others. The series were cohesive throughout, showing two distinctive features: "sensitivity" and "fragmented transitional object" in the separation series. Sensitivity means that all three PORT series during the exposures 1–3 are reported essentially as "correct" from the point of view of their content – an adult and a child in the picture, for example, in the separation series where the woman has her back to the child. It does not rule out that distortions may occur later in the series. Sensitivity in PORT is interpreted as an unusual ability of the patient to unconsciously/recognitively read subtle expressions in other people and "directly" get past the affect of surprise (which is missing) – directly towards "recognitive fear/dissmell/anger"? – as a "gradient affecting preparedness". As already mentioned, a specific answer is "fragmented transitional object", exemplified in Figure 6.9. The patient in the example is a thirty-nine-year-old man, diagnosed with psychopathy and sentenced to psychiatric inpatient care for attempted murder.

Three of the patients reported the transitional object to be "a marble", and it was then reported by one of the patients to be "a marble split into a thousand bits" in the following exposure. "Marble" as transitional object has so far only occurred in patients with a diagnosis of psychopathy.

Exp. 16 'It's a doll.'

Exp. 17 'It's a doll broken into pieces.'

Figure 6.9. Examples of "transitional object" (exposure 16) and "fragmented transitional object" (exposure 17). The separation theme.

How should one understand and interpret the fragmented transitional object in the separation theme? If we follow Winnicott, the transitional object has an important function of bridging the relationship between "me" and "not-me" and the relationship between the world of Man and the outer world of physical things/manufactured objects/artefacts. In view of the fact that from the beginning in the potential world of the psychopath it is a matter of early triggered fear-terror, in its turn overlaid with *deprivation* of the affect of interest – what type of "morality" does one have to stick to when tracing authentic interest (which psychopaths are insensitive to) in the human world? Everything in the transition "me-other" and "human-physical world" is fragmented, "ragged". Revenge and retaliation become a latent emotive script. When acting as a psychotherapist, on one occasion I traced in a client dissmell in collusion with anger as emotion, and after months did not feel the least hint of either interest or attachment/transference in the client, I suggested more conversations/week. The client refused and failed to show up at the next session. Upon contact with him, despite the offer of a new appointment, he refused and we agreed to terminate the therapy.

I have never conducted any systematic and controlled study of patients diagnosed with "narcissistic personality disorder". For many years I was in contact with the staff at a treatment clinic for addicts, now closed. The treatment approach included both departmental and individual treatment on a psychodynamic basis. All patients accepted were tested with PORT. My task was to blindly code and blindly interpret the PORT protocol. This meant that I was not informed about the diagnosis until afterwards. The diagnosis, according to DSM, was performed by a psychiatrist attached to the treatment clinic. The last exposure (20) for a twenty-two-year-old male patient, with "narcissistic personality disorder" is presented in Figure 6.10. As appears in the figure, the patient does not achieve a correct report of the child in the picture. The woman/mother is reported to be holding a mirror in her hand, according to the manual coded as a lack of attachment relationship (c). Afterwards, I had access to the patient's records. In it, there was a detailed account of the patient's psychical history from early childhood. The young man had grown up with a single mother. This did not exclude that several men had passed by in the environment where he grew up. The journal also revealed that the mother repeatedly tied the patient when he was a little child (during visits by men when she worked as a "luxury prostitute") with a string attached to a radiator. Is there any way to trace this in the PORT series? I do not think it can be better illustrated than in the excerpts retrieved from the separation series (Figure 6.11).

Exp. 20

'She is sitting with the right leg over the left. Holding still a mirror.'

Figure 6.10. The last exposure (20) for a male patient with a diagnosis of "narcissistic personality disorder". The attachment theme.

According to the manual, the non-existing "ribbon" (often referred to as a "leash") is coded in the separation theme as "secondary dependency". Do we know anything about the patient's mother? She visited her son at the treatment clinic on various occasions. Several persons in the staff had met her. The overall description of her was that nothing indicated substance abuse problems (but "did she abuse men?"). She was described as being "cold, stylish and narcissistically omnipotent".

Exp. 6

'I think I saw a dog held ... with a leash.
Somebody in the doorway pulling.

Exp. 12

'Yes, exactly the same.
But that creature probably a child.
Was a little bit higher up.'

Figure 6.11. Exposures 6 and 12 in the separation theme for a male patient with a diagnosis of "narcissistic personality disorder". Scored "secondary dependency as warding off".

In a comprehensive study, five groups of LVU juveniles aged 16–18 were studied, they had been tested with PORT during the 1980s and 1990s. In brief, the young people tested during the 1990s showed PORT patterns that largely overlapped with those of the borderline patients (tested during the 1980s). The youth in the 1980s showed, in particular, signs of "depression (a-d)" – serious signs in themselves, but not of the same deep pathological dignity as the PORT pattern that the juveniles of the 1990s showed.[45] As illustrations I use two sequences, one of them retrieved from Martin's attachment series (Figure 6.12), the other one from Adam's separation series (Figure 6.13) (both Martin and Adam were part of the 1990s groups). Martin's problem was that he had attempted rape, always women older than himself. Figure 6.12 shows three exposures from his attachment-series.

In exposure 7 (and in previous exposures) Martin reports and Draws a face scored as "whole configurations as warding off". In exposure 19, he reports a woman and a baby, largely a correct representation of the picture shown. Elements not considered to be correct in the coding are that the child is "standing up" and "A dirty wall" ("depression c"). Martin's report in exposure 20 is remarkable: a woman and a man aged thirty, with the man straddling the knees of a woman, a sort of sexual intercourse position (a psychoanalyst may have noted that Martin draws the reported man on the picture in the form of a phallus). The coding in the attachment theme is "sexualization (a)", responses in PORT that I saw for the first time in a group of patients sentenced to inpatient psychiatric care for repeated rapes.

A detailed account of Adam's psychopathological history, symptoms and treatment at the LVU clinic where he was admitted are included in my second book in Swedish. From early school age and throughout the teen-age period, there was a continuous theme of serious abuse of alcohol – his father was a criminal and an alcoholic – the main reason for compulsory inpatient care at the LVU clinic. In Figure 6.13 Adam's last four reports in the separation series are revealed.

45 E. da Silva, S. Ellbin, E. Marne, A. Nilsson and B. Svensson, 1999.

Exp. 7

'Still a face. Light around.'

Exp. 19

'Looks like the baby is standing on her knee. Plank floor. A dirty wall.'

Exp. 20

'A carpet under the chair. Another person, a man is straddling the woman. Both about thirty years old.'

Figure 6.12. Excerpt from LVU juvenile Martin's attachment series.

Exp. 17

'A woman with a dress,
the child a nappy.
The woman is going into
another room.'

Exp. 18

'No things in the room.
Bare room.
No pictures.
Wood floor.'

Exp. 19

'The kid wants to have what she has in her hand. She is standing with her back to the room. She has a teddy bear or a feeding bottle in her hand.'

Exp. 20

'Yes, this picture. It's clearer now. She puts the kid away with the help of a bottle in her hand.'

Figure 6.13. The last sequence from the LVU juvenile Adam's separation theme.

In exposures 17 and 18, the door in the drawing is shifted downwards to the bottom line of the displayed picture – “splitting (a)”. In exposure 19 there is no configuration of splitting. Instead, the child and the woman/mother unite through a highly concrete transitional object in the picture: “a teddy bear or a feeding bottle”. In exposure 20 splitting is back in the drawing, the woman on one side with the feeding bottle in her hand, the child on the other side “deprived of the access to the desirable feeding bottle” as the “oral” transitional object. One may surely assume that in the depths (*Srip*) of Adam it concerns *deprivation* – with “bottle” as the only comfort and rescue.

Figure 6.14 illustrates how depression (a) is scored in PORT’s separation series. The most serious variant (a) is coded when the patient reports the child and/or the woman as being “dead”, “a corpse” etc. The illustration comes from a report by a fifty-three-year-old woman, admitted to a psychiatric clinic and treated for severe depression, sometimes interrupted by short relapses of mania.

Exp. 9

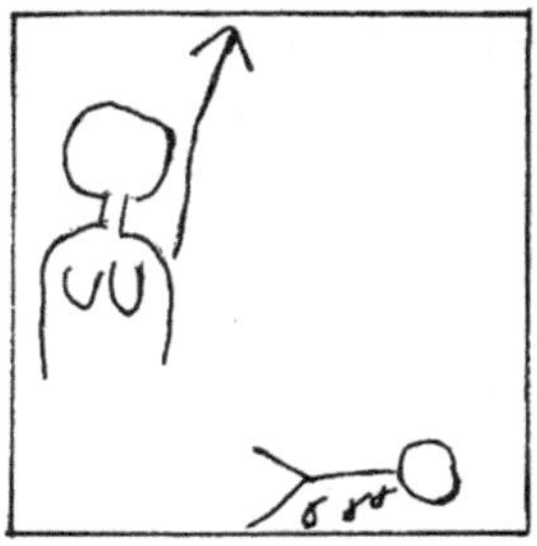

‘It’s a dead human bleeding.
And the woman who is
going away.
No strength to draw.’

Figure 6.14. Illustration of scored “depression (a)”. The separation theme.

Some words about the most recently published research with PORT as the survey method. In 2011, a second repeated study (a cross-validation study) was published on “suicidal” patients. In 2004, the first one was published. In both studies, with two different groups of suicidal patients, a uniform cluster appeared in the PORT series of the patients: abundant elements of “dissolution anxiety (a)”, “lack of attachment relationship (a-c)” (attachment theme) and in the Oedipal series, there were elements of "motoric activity/acting out".[46]

As a reader, you might wonder how repression/suppression and isolation, in the form of wardings off are expressed in the PORT. Repression and isolation are beyond the scope of signs that can be coded in PORT. For the operationalization of these two wardings off, I refer to Ulf Kragh's method DMT, *Defense Mechanism Test*.[47]

Space, synchronous/now/Augenblick, simultaneity, moment, continuity

It is worth repeating: there is something special about the visual sense in relation to other sensory systems. This applies from the beginning: the infant in mutual eye contact with the mothering world with tunings in and tunings out of the innate affect. But then also throughout the continuation of ontogenesis: the visual sense and the visual perception that become superordinate to all other sensory modalities (Lev Vygotsky). As with no other sense, the domain of vision is space – with time as a hidden reverse side. Then it becomes understandable that the infant until school age has a visual perception that largely overlaps with the child's consciousness and where the organization of time of what has been is extremely inadequate (Alison Gopnik). Moreover, there is the absence of inner silent speech to come to terms with when in the direction of the time-arrow, one thing occurs/occurred/will occur after the other in relation to the other ... (Teresa Bejarano). For the infant, "*where*" (somewhere) is thus a step ahead of "*when*" (something happened) – which simultaneously means that the analogue affect/the emotive scripts and perception are still a step "above

46 D. Titelman, A. Nilsson, J. Estari and D. Wasserman, 2004; D. Titelman, A. Nilsson, B. Svensson, H. Karlsson and S. Bruchfeld, 2011.
47 Ulf Kragh, 1968.

in command" in relation to the digital cognitive operation. I take Anthony Wilden at his word: "The analogue on the other hand talks only about relationships."[48]

Can this be understood in the evolutionary perspective? Simply put, the *place* for the exchange (which still applies in the world of animals) has taken precedence at the expense of the *occasion*. Following the prey in its footsteps may in the world of *Homo* have become a concern of sight and seeing at the expense of the power of other senses. The smell that initially puts the innate child on the tracks, during the course of ontogenesis, is assigned an increasingly *secluded* place – in adult life increasingly sliding towards an either-or luxury character: the good smells to whet one's appetite, the sinister odours to warn us about the rotten food's decomposition. Could this also be regarded as a clue to the auxiliary and relation affect dissmell's characteristic of either-or?[49]

If you follow my description from the beginning in this chapter, there is no evoked internal presentation without reference in a recognitive emotive value configuration. As I explained earlier, the choice of picture content to be displayed is of crucial importance. Thus, recent research has shown that subliminal display of a picture showing a happy face results in increased activity in the smiling muscle, *zygomaticus*; the picture of an angry face gives rise to increased activity in the frowning muscles, *corrugator*.[50]

DMT includes pictures showing two persons: father-son and mother-daughter pictures, with the father/mother in the background of the picture. The face of the background person expresses anger-rage, almost in a grotesque manner, to signal threat/danger. Without doubt, by displaying pictures of faces/persons one may subliminally control the mood of the person who is exposed to these – albeit unconsciously/recognitively for the person him- or herself. Ultimately, it is possible to refer to the contagiousness of the affect, translated into facial expression, to explain the phenomenon.

48 Anthony Wilden, 1972/1975, p. 164.

49 Silvan Tomkins, 1991.

50 Ulf Dimberg, 1982; U. Dimberg, M. Thunberg and K. Elmehed, 2000; Marianne Sonnby-Borgström, 2002.

From the evolutionary perspective, I will argue that with the emergence of the mammals and caring, an elementary affect-sensory-internal presentation system for reading the affective expressions of the mother and the offspring probably had a decisive adaptive value. Recognition of similarities relative to dissimilarities became an important element in the context of feeding and caring. For the offspring, to recognize the mothering world should give a clear (affective) signal of (feeding) exchange; lack of (affective) recognition, a signal of danger of becoming – a prey. If it once started this way, it has been along this way things continued, all the way to the nine human innate affects, transformed into emotive signals/visual internal presentations in scripts for recognition in the adaptively functional *Srip*. During the development of the mammals' flock existence, towards a life in an increasingly complex *social* world, *Srip* continued to have an adaptive and functional value: primarily for the recognition of emotive scripts translated into relationships (according to the analogue principle). It was, and still is, in the relationships that the value indications emerge: on the one hand of positive value, the positive affects, on the other hand of negative value, the negative affects. In parallel, the fact that I have been calling "the premature birth of the human child" occurred; further, the long childhood period with reference to, and dependence on, the parent world. Remember that it is not until puberty that sexual maturity occurs.

A picture is a picture, is a picture, is ... It applies to the visual perceptual world. To release the influence of the operative cognition over perception in PORT, short-term exposures (from a subliminal level) of pictures are used. As stated, there is no sense that is more prominent from birth than sight to read affective communication. It is with the ability for evocative internal presentation transformed into being in service of the visual system that we can safely argue that visual perception has reached the heights that prevail in adulthood. There are two questions left to try to find the answers to in order to be able to explain why reported PORT illustrations probably have the value/meaning I attribute to them in my comments above. The first question is why it is primarily psychopathogenic expressions that are interpretable in a meaningful manner.

In the PORT theme drawings, the persons are deliberately presented with neutral facial expressions. Perceiving danger-threat in the pictures probably has to do with experiences that relate to the attachment phase and/or the separation/individuation/autonomic phase and/or "Oedipal/Electra"/gender phase (and all experiences with reference in these phases later in life). With the contingency S-R-SE and the affects as the first communicative "language", as a starting point, it is probably quite obvious: excess of *frustration* (triggered by the affect of fear) and/or *deprivation* (triggered by the affect of distress) and/or *irritation* (triggered by the affect anger) make imprints in the form of *traumas*. There is a saying, "health is silent" – in the PORT genesis without significant deviations in the construction of the picture content which is shown. The construction is normally built by adding details to details, stepwise. Trauma always causes scars – in psychopathologies, with a metaphor, with the scabs remaining – it applies to the development of soma as well as to development of the psyche. Scars/wounds are visible – that's exactly what we make use of in PORT.

The second question is about percept-genesis and continuity. For attempts to answer that question, I refer to previous arguments which point out that "simultaneity" takes time (books four and five in Swedish). In my description, "simultaneity" is made up of a series of "recognitive synchronous/now/Augenblick" – and the series requires approximately 0.5 seconds to reach "simultaneity" (Nicholas Humphrey, Benjamin Libet). Within that timespan there is room for exactly fourteen exposures or "recognitive synchronous/now/Augenblick" (exposure fifteen takes exactly 0.5 seconds), in a PORT test.[51] It has already been found that patients rarely achieve "simultaneity" – the exposed picture still has features of the "recognitive synchronous/ now/Augenblick" – thus rarely achieves the correct perception of the displayed PORT picture, even at the exposure twenty which is displayed for two seconds. And the explanation for it would probably be that they have a large number of constraints/formation of wardings off to pass by/through (within themselves) in order to come to grips with, for example, a traumatized "separation/individuation/autonomy theme". As long as a picture is a picture, is a picture ... exclusively in the domain of visual perception, *recognitive emotive value configurations* influence the

51 Alf Nilsson, 2009, chapter 3.

evocative internal presentations. That is what the patient is told to convey in the percept-genetic test situation. And the background is that the onefold positive and negative affects are without "operative access" and perception, at most, manages time diachronously, in the sense that it never devotes any *attention* to time (but only triggers surprise-startle, but not expectation). Therefore, affect → emotion → feeling always take the time they need – and never end up lacking time. Briefly: affect → emotion → feeling work synchronically → diachronously, the latter most so, and only care about relations, according to the analogue order (Anthony Wilden).

A small addition to the question of what the dynamics of the PORT progresses consist of. For the person without psychopathological overlays, similarities remain across the exposures – but perhaps also some dissimilarities in the details. Each reduction of dissimilarity increases the degree of similarity: one picture is the same picture, is the same picture, is ... until the C-phase, correct perception. In this person, *the constraints* in the form of *wardings off* as "potentials" are not sufficient to be triggered by a "picture". Instead, the cognitive operation is still in charge, a "picture" leads to a thought, possible to translate into language. A constraint can never fully be explained. For the constraint retains potentials in a system, or in an organization, or in a pattern. The potential is, as already mentioned, nothing present, instead it is something "absent" – and thus not possible to *make fully visible*.[52] How different it probably is, then, for the patient with psychopathological overlays. As long as the patient remains in the pattern of excessive negative emotional scripts linked to wardings off, for example, the traumatic separation experience within "me" is a reality, is a reality, is ... At the level of recognitive internal presentations, a patient who has these problems of separation to cope with is reminded of these problems until the level of evocative internal presentations is prepared to accept the experience of these problems as "only an outer picture of them". This would be at the *moment* when the diachronous perception would have "convinced" the asynchronous operative cognition that this is only a picture, is only a picture, is ...

52 Terrence Deacon, 2012, chapter 6.

CHAPTER 7

Return to the start: and the whole extension to *Homo psychicus* today

> What happens within me happens, I sense beyond what feels like myself in me.[1]

Introduction

The epilogue that introduces this book was formulated by Frank Close.[2] His book is an excellent compilation of what Man knows about the creation of the universe (Big Bang), through "the entry of mass that gave substance and form to the universe" and all the way up to the finding of the existence of the Higgs particle, the boson. Moreover, one may add, "mass is the spoiler of the symmetry of the Creation and the source of all structure, pattern, and asymmetries since".[3] Perhaps you wonder: "What does this have to do with the human being?" I will let Close speak once more. It is:

> carbon, nitrogen, and oxygen [...] that are the templates of amino acids and life. Collect enough atoms of these elements together and they become self-aware. [...] [W]e do know that it takes a considerable amount of time: over four billions years of sunlight shining on the earth was needed before intelligence at the human scale emerged.[4]

1 Alf Nilsson, 2018.
2 Frank Close, 2000/2013, p. 256.
3 Frank Close, 2000/2013, p. 240.
4 Frank Close, 2000/2013, p. 244.

Given the temporal perspective Close conveys, one can hardly feel anything else than smallness, and above all humility. What is Man about to stir up in the biosphere, with consequences for living conditions on earth?

Homo psychicus: early in evolution and phylogenesis

The beginning, and the prerequisite for the emergence of *Homo psychicus* during evolution, was the genesis of mammals. What benefit did evolution bring with the innovation of the mammals? The big advantage that I can see is that with the diet of mammalian child, another step was secured on the scale of survival with the mother's nutritional circulation inside, and thus the mother as an ever-present nutritional depot for the child. Also, it gave the mother the possibility to constantly keep an eye on the child and his or her well-being, and to provide warmth and protection, in Sarah Blaffer Hrdy's words, to offer "care" (at the expense of the male/father's task of providing the mother with the necessary supplement of food for her to eat). Moreover, the necessary body contact between the child and the mother can be added, in order to keep the child alive through suckling. From this you can derive imprinting, binding, attachment, dependency: transformed to such a degree and extent in connection with the human child – beyond what applies to every other living being.

Now I will take quick and big step over to the issue of the origin of *Homo* – and further all the way to *Homo psychicus* as we know him and know about him today. A decisive transition to appropriate, according to the approach of this book, is the foundation of the gradient affect of surprise-startle, expectation a priori and expectation a posteriori as well as spread-overs and interactions between them. This I am inclined to designate as to the basic infrastructure of *Homo Psychicus*. To Silvan Tomkins' general derivation of the nine innate affects as amplifying stimulus as well as response, I will make an exception for the "neutral" surprise, as merely a "reactive" stimulus amplifier, with a focus on always taking the first command.

There are thus exceptions, as I show with the example in Chapter 4, to surprise, usually, primarily to trigger the eight other "value effects" on the line of affect/internal presentation (the affects as the "primary motivation system") and expectation a priori and a posteriori on the operative line of development. There is obviously an element of "autonomy" regarding expectation with all its extension in the development towards operative cognition/reflecting abstraction. All in all, therefore, the relationship between surprise and expectation (a priori and a posteriori) seems complex and does not lend itself to being easily defined. But, as a basic rule for human life, it is true that without the nutrition of the affects, operative cognition idles. Before moving on with surprise in relation to expectation – let us dwell for a while on expectation a priori versus a posteriori.

It is clear from the above that in this distinction I follow Teresa Bejarano. Nowhere does she make any reference to Immanuel Kant's epistemological thinking about "a priori knowledge" ("prior knowledge", independent of outer expressions) versus "a posteriori knowledge" ("after knowledge", after outer impressions).[5] I accept Bejarano's definition, which I have adopted as my own, of (i) a priori expectation, that it may well be included as the basic element on the operative developmental line and already then in the innate reflexes, and (ii) a posteriori expectation as the necessary element – with its reference in the mimicking of the (care)givers' kinaesthetic and sensorimotor expressions – and as my contribution, for dealing with physical things/manufactured objects/artefacts. The consequence of this is that imitation in order to learn – and learning in the broadest sense as it is manifested in ontogenesis – originates from expectation a posteriori. Thus, expectation a posteriori is the necessary prerequisite for the child's ability, through "deductive learning", to quickly adapt in order to deal with physical things/manufactured objects/artefacts (in the long run with the ability to *classify* them). With this, Man has secured something that is unique to him: that with mimicking beyond that on the immediate level of the mirror neuron, along with the addition of expectation a posteriori, according to my proposal, only reserved for him, achieves the ability to

5 Immanuel Kant, 1771/2004.

imitate as a way of learning, in its turn with "deductive learning" as an extension.[6] But in order to reach all the way to this, the "world view" of physical things/manufactured objects/artefacts stretched out towards the cosmos, an extended childhood is needed up to adolescence, when reflecting abstraction reaches the child (Piaget). Moreover, to this is added the need to have access to language.

After this attempt to clarify expectation, let us return to the absolute origin of surprise, expectation a priori and expectation a posteriori. At the risk of repetition: surprise is a basic element of the innate motivation order, the affect system; expectation a priori and a posteriori, the basic elements of the innate operation system, with mimicking and the set of reflexes as the first available. The common feature of surprise (in the world of protected people mostly) and expectation a posteriori is that they originally have reference in the caring world/other person. But there is also an absolute difference between them: surprise primarily leads further into the "value affects" and via them further on with the reflections in the world of human relations; expectation a posteriori via the caring world/other's kinaesthetic body position and expression of action directly towards the world of physical things/manufactured objects/artefacts.

Affects are onefold, analogue and continuous. Operations are twofold, digital/discrete and discontinuous. Affects inserted into the evolutionary perspective are "late" with the emergence of the mammals; operations "early" with the emergence of the vertebrate animals. The ultimate foundation of the operation I have anchored in the rhythmic activity/biphasic cycle. My suggestion is now that during the phylogenesis of *Homo sapiens* – as everything suggests, via a first "technological explosion of a kind not seen before" (Richard Leakey), about 50,000 years ago – the twofold was transformed into on the one hand expectation a priori, on the other hand into expectation a posteriori. The next big step in this direction took place around 10,000 years ago, when humans embraced agriculture, cattle raising and sedentary habitation. Since then, the road has been open to

6 In Alf Nilsson, 2013, pp. 61–62, I present examples of how to get chimpanzees, born, nurtured and raised by humans, to mimic humans' facial expressions (stretching out their tongue, pouting) like innate human children, for a limited period (5–11 weeks).

When the "space" of *Homo sapiens* was filled with more and more ıysical things/manufactured objects/artefacts, then expectation a pos-eriori probably developed to the extent that applies to us humans today. ın this context, it would be assumed that its scope is less open to another's area of action as such, but is rather directed towards the physical things/manufactured objects/artefacts that the human world is filled with. There, expectation a posteriori came in handy as a shortcut to the swelling space of physical things/manufactured objects/artefacts without the affect of surprise-startle being in need of triggers, whether appropriate or not. But: no *discovery* of physical things/circumstances and no *innovation* of manufactured objects and/or artefacts without taking the road through the triggering of surprise and above all *interest* – to which I would like to add "to some excess". More parenthetical in this connection is my suggestion that a measure of "habit" in humans applies to how many times surprise and interest along with expectation a posteriori have been screened already, something that has determined the depth of "habit".

Before moving on with the dynamics between expectation (a priori, a posteriori) and surprise – a little additional space should be devoted to operation. Actually, I would only need to refer to one argument for its existence as a feature in the equipment of *Homo psychicus*: Man would never have been able to invent the digital machine/the computer as something external, unless the prerequisites within himself were there first as something inner, in order to stage it in the outer. But let's take the matter one step further in the perspective of space-time. Helmar Frank, early on, coined the term "subjectives Zeitquantum, SZQ", a kind of minimum measure on our time-dissolving ability. Frank translated this into "one bit per SZQ".[10] With the precision instruments available today, we can ignore Frank's quantitative definition – but not the principle of space-time that he was looking for.

In life on earth, we live, with perception and cognition as yardsticks, in a "field" with four dimensions, one time-dimension and three space-dimensions. As adults, we live in our everyday life until death in diachronous

10 Helmar Frank, 1962.

time-dimension. So, it still seems to be, in the entire animal world beyond human beings, as animals. It is not uncommon today to refer to bipedalism as a breaking point for *Homo* – and with that we could fixate our eyes and see our hands. With that followed, as *one* extension, among other things, the view of space as "infinitive redundancy". In the same synchronous/now/Augenblick, it became clear to Man (beyond the animal) that besides the physical things there was "infinite redundancy" for him to fill in with manufactured objects and artefacts. With what we know today, it is an illusion that we, despite postmodernity, tend to linger. After postmodernism's erasure of the boundaries between the sign and the sign's reference, followed the erasing of the boundaries between economics and politics. We are still living in the post-political era. Already in the early 1990s there would probably have been some indications of it.[11] The guiding principle for this economism is still the illusion of diachronous time and the redundancy of space as infinite. As it seems today, we live in the hands of an economic-political elite alienated from the humans with their long (evolutionary) history. In this flow, the thought that there are constraints along the way to consider is not embraced. Can this be attributed solely to the innovation of the digital machine and cyberspace? To answer the question, I will take a big step back in time. Yes, back to the time to when the innate action operation crossed over to the constraint/interface and into the cognitive domain. The question at this point in time that I think would be worth asking is: How should we use our so well-sharpened operative cognitive ability in order to serve us as human beings, ultimately for life on earth? Thereafter, a recurrence of my previous proposal followed by a question: If operative cognition has borrowed traits from the most *abstract* of the negative affects, the analogue *level*, and thereby "extended" *anger-rage*: Is operative cognition unstoppable before the complete destruction and the absolute extinction of life on earth? To return to the time we live in now: It doesn't work anymore, after Alan Turing's discovery of the "halting problem" of the universal Turing machine, to refer to: "This is nothing we knew about!" The halting problem in a nutshell: "There is no algorithm to

11 Alf Nilsson, 1992.

decide whether a Turing machine will ever stop". Protagoras already knew this. And it became extremely firm with Albert Einstein's brilliant formula for space-time in the universe: $E = mc^2$.

Somewhat schematically, Man had at this time, with access to operative cognition, created "a square for time" capable of being combined with and subordinate to "a square for space". Hence, the human being had placed himself outside – from above? – the course of diachronous time on earth, and the life there going back about four billion years in time.

Neither can you refer to lack of insight about Man as an affective, emotional and feeling being. Charles Darwin, as is made clear, already put us on the track. Feeling is the first thing we are aware of in all of our experiences. One question worth asking is why we let cognition at the expense of the affects get such an academic supremacy. The twofold and overweening confidence of cognition against the onefold and underestimation of affect(ion)?

With that, back to the dynamics between surprise, expectation a priori and expectation a posteriori. In the case of expectation a priori, it remains on the operative line of development, all the way towards our ability for "a priori knowledge" (Kant), a concern for the reflecting abstraction. What remains to be followed in development is surprise-startle and expectation a posteriori. My suggestion in Chapter 3 is that the biological affects reach a first *psychical* conversion in *Srip*. On this, in adult life, recognitive/unconscious level – that which lies beyond evocation cannot become an element of our consciousness – affects converted into emotions have dominion over "the familiar" (face) in relation to the "strange" (face). This should always apply in ontogenesis up to the acquisition of multimodal evocative ability, with the enclosing visual evocation being superior to other sensualities. We have derived expectation a posteriori and mimicking via the prerequisites of the caregivers'/others' stretching out towards the physical things/manufactured objects/artefacts. My suggestion is now that surprise and expectation a posteriori, both with the deepest reference in the caregivers/other, follow each other closely: On the one hand a posteriori expectation ↔ *surprise*, all the way from triggering of the value affects, translated into emotions, towards the evoked feelings, with the whole line that has the contingency S-R-SE as a pattern formation. In this we can attribute to epigenesis the decisive influence on the adult person as a moral, ideological, societal and social being. As shown in Chapter 4, this analogue development takes place with emergences of the motivation orders: first desire and later yearning as interfaces.

On the other hand, surprise ↔ *expectation* a posteriori with the development of formal cognitive operations, through primary and concrete cognitive operations, primarily the pattern formation (with the help of language) with regard to physical things/manufactured objects/artefacts, all the way up to the conversion into *classes*. Right in "the silent" within us, *expectation* a posteriori becomes the bridge between the evocative images, with their *categories*, and the operatively functioning reflecting abstraction. How is it then with reflecting abstraction in relation to positive and negative effects/emotions/feelings – the whole way to a standstill? Now I'm inclined to attribute the making of "me-ness into a whole" to the evoked feelings, that within me which feels like myself in me. It is usually our common standstill for the expressions of "me-ness" – but not without exceptions, Sigmund Freud being one of these examples of an exception, with his abilities to self-analysis. But to this comes language, without which we would not have been conscious of being conscious – and in addition, speechless. I will return to language, but first a few more comments on percept-genesis.

Percept-genesis and visual perception once more

In my fourth book in Swedish I refer to Anthony Wilden's thesis about the analogous continuum of visual perception's "more-or-less presence-absence". Can it be understood, if we consider sight as a modality? Is it not so, that either a thing is within the field of vision and we perceive it as being present, or it is outside the field of vision and we perceive it as absent? Nowadays, we know that it is not that simple. We may have something in the field of vision that we do not see – but still perceive subliminally. That we perceive visual stimuli subliminally can be described as an unconscious *readiness* for the perception of visual stimuli. Is it possible to write about what I call in this context "unconscious/recognitive readiness", in terms of Wilden's continuum of presence-absence? I will give it a try.

In the analogous functional visual perception, there is a constant *readiness* for presence, or does the presence linger in order to *continuously* attach to the absence within us – the absence as it appeared at the synchronous/now/Augenblick for the recent presence of a thing in the visual field, and at the synchronous/now/Augenblick just when the thing ended up leaving it, in absence. Described in this way, it may also be perceived as a contribution to the understanding of the "continuity", or the silent/invisible diachronous time in visual perception. It is in line with the previous description of the gaze not being able to get a hold of time, but only of things and events in space (as they occur diachronously). To this I would first like to add a formulation from Teresa Bejarano, which hopefully helps clarify my own description: "Negative predicates cannot come from perceptive data [...]. Perception could be incomplete but never negative." In addition, a formulation from Wilden that I perceive as being related to Bejarano's: "In the analog computer [like analog perception], one cannot say 'not'".[12]

The most ingenious feature of Alan Turing's discovery of the code for the digital machine I've described in terms of his being able to "deceive" the machine on time by transforming time into "a sequence of events in space". To do that, a minimum unit of information in the form of bits was needed. What is it that we are doing in percept-genesis when we try to "deceive" perception on diachronous time, and do not give it "the usual time for simultaneity", 0.5 seconds, but only give it "a small fraction of time" – in the form of what I refer to in Chapter 6 as "recognitive synchronous/now/Augenblick" – to perceive the displayed picture? Yes, we try to "deceive" visual perception on the diachronous time that is needed for perception of "simultaneity" by allowing a display of the pictorial theme in "13 recognitive synchronous/now/Augenblick" sequentially. For all 13 "recognitive synchronous/now/ ...", the display time is less than 0.5 seconds (as shown in Chapter 6, the first exposure lasts for 0.01 seconds). Can this be set against what applies to the digital machine? If I put it as follows:

12 Teresa Bejarano, 2011, pp. 222–223; Anthony Wilden, 1972/1977, p. 24.

In "the universal Turing machine", the numbers of mathematics that "do something" are used – they transform time into "a sequence of events in space" by utilizing electronics/feedback circuits that do this during at most a "time gap", which causes the machine to neglect it as being time. In the "percept-genetic machine", the tachistoscope, with the ability to add more and more (time), one "deceives" perception on the diachronous time needed for conscious "simultaneity" by breaking it down into "recognitive synchronous/now/Augenblick" – but with enough duration for *Srip*/unconscious experience of the picture content (the *Srip* element in perception). Ulf Kragh's and Gudmund Smith's stroke of genius was to show, from a subliminal level, picture themes with person dyads according to the contingency S-R-SE – picture themes that touch more or less general human experiences – one and the same picture in a continuous sequence for reconstruction/evocation. PORT includes picture themes that refer to the generic experiences of attachment, separation/individuation/autonomy and the beginning of gender identity. With reference to the two-layer theory of perception reported in this book, the person/patient in the PORT situation is asked to draw and verbally explain the *evoked* internal presentation – immediately coloured by the "value icon" triggered in *Srip* – immediately after the exposure of the picture theme. The fact that in patients who have experience of traumas containing elements of excessive negative affects/emotional scripts, "constraints" are raised → "wardings off" or "affective exertions", may perhaps not seem to the reader to be utterly unreasonable.

Is there any neuroscientific knowledge to support my reasoning? There are studies that show:

> that any emotionally charged visual image is encoded [...] and stored in the same higher-order areas of the visual system [in the cortex] that initially processed the information about the image.

To that is added: "our brain processes aspects of the image that relate to emotional perception more rapidly than aspects that relate to form [with a difference of 100 milliseconds]."[13]

Moreover, PORT is anchored in extensive clinical-empirical evidenced-based research.

Language as a sign: from icon via index to symbol

Is the word, with the undeniable extension language, an *epiphany*, a luminous sprig on the branch of evolution with Man at the very top of the branches? Or is language a *fragment* of incompleteness, that fascinates the listener (viewer) intending to complete it? If we follow Charles Sanders Peirce (see above), language is a phylogenetic success story rooted in the icon, based on similarity, the index based on proximity (in time and space) and the symbol, based on rules. Among the given introductory keys in Chapter 1, which semiotics/linguistics have coined and reserved for the specification of language, there are deixis/deictics, predicative and theme-rheme. My hope is that you have comprehended that these terms are equally valid for the general description of Man as both an affective/internal presentational and operative being. Epiphany or fragment? My answer to the question is both. It is phylogenesis's most perfected – paradoxical? – combination of the analogue and the digital into a whole.

Actually, I have probably conveyed most of what I have to say about language in Chapter 5. In spite of that, for heuristic purposes, I will take another round with what I perceive to be the most difficult element of Peirce's linguistic step hierarchy, namely, the origin of language in the icon.

The icon is based on similarity. As such, it builds on the analogous principle: one thing that resembles another and the things that follow each other in the diachronous time. In this book's systemic approach, this applies

13 Eric Kandel, 2012, p. 308, p. 345.

to the "logic" of affect and perception. As I see it, it is to this "logic" we need to direct our gaze – or direct our hearing? – to track the roots of the icon.

In my first book in Swedish about *Homo psychicus*, I describe the "bandwidth" of the various sense modalities, that is, "how many bits per second" they can include, convey and make way for within us (Benjamin Libet). Sight has the largest bandwidth of ten million bits per second; hearing about 100,000 bits per second. This is based on studies of adult individuals. If one presents light pulses of pictures with a frequency of less than 16 per second, one perceives it as "flickering pictures"; at a higher frequency as "*moving images*". Corresponding to audio pulses is the perception of "pulses", and *a "tone"*, respectively. Can these "brute facts" give us some guidance in our understanding of what an icon fundamentally is? My only conclusion is that there is a fundamental difference between vision and hearing as modalities. Can we speak about a tone that is similar to a tone? Two tones with an octave between them (without having absolute pitch) we perceive to be the *same tone* and not a tone similar to another. For sight, getting a glimpse of a real pipe and the picture of the same pipe, the pipes *resemble* each other. At the same time, we *know* that the real pipe is something other than the picture of the pipe. For the hearing to perceive a melody or a language that resembles another melody or language, it will not suffice with a single glimpse. For that, *time* is needed. For sight, in order for it to perceive similarities, it does not have to cross differences – only, at most, dissimilarities? Does hearing, on the other hand need at least two points of reference – difference and similarity – to distinguish similarity? Had we never *heard* and perceived the difference between a, b, c ... we would never have had a finite field for language. Let us play with the idea that we had never heard the difference between a, b, c ... but had only been exposed to the letters visually. My thought is that, as both a glimpse and over time, we would have ended up looking for similarities between the letters – but would never have considered the differences between them and used them as such. *There probably is a fundamental difference between vision and hearing as sense modalities*. Are there any neurobiological bases, then, for any difference that is suggested here? In quite a lengthy interlude, I will let Jennifer Groh become my spokesperson to give some possible evidence of this.

An interlude with Jennifer Groh as spokesperson

Jennifer Groh differentiates between two forms of brain encodings:

> In a map, the *locus* of neural activity represents information, but in a meter, the *amount* of activity represents information. […] The representations used by computers […] come in two flavors, digital and analog, paralleling the brain's maps and meters. […] The brain's *maps* of information can be thought of as a neural version of a *digital code* (thus, the brain's *meters* of information as a neural version of an *analog* code). [the latter italics and parentheses mine] […] If a given neuron is firing in the brain's visual maps, this signals the presence of a visual stimulus at a particular location in the visual scene. […] Of, course, this is an oversimplification, as neurons actually have graded, not binary, responses. That is, neurons are not simply "on" or "off" but can generate widely varying numbers of action potentials. This property resembles analog, rather than digital, signaling. […] Thus, any given populations of neurons can *simultaneously* participate in two types of code: which neurons are on and how strongly they are on. […]. When a (metered) movement is guided by a (mapped) sensory event, such as *when you point to something you see* [remind yourself as a reader of Jonathan in Chapter 3] [my italics and brackets] the brain must translate between these two internal languages. Such a map-to-meter conversion is the brain's equivalent of digital-to-analog conversion. […] The digital-to-analog conversion involved in visual guidance of eye movements is thought to rely on a brain structure known as *superior colliculus*. […] There are two sets of mounds on the top of the brain stem […]. The lower pair, the *inferior colliculi*, are involved in hearing […]. The upper pair, the superior colliculi, are involved in controlling where the eyes are looking. […] The superior colliculus serves as a *bridge* between *maps* employed by the visual system and the *meters* used in the motor system [the eye has six distinct muscles for moving it, the only sensory system moved by muscles]. [the last three italics and parentheses mine] […] [T]he brain [is not] a handy-dandy guide to sound directly from the ear, the way the visual and somatosensory systems do from their sensory receptors. Instead, it has to build one. So there are some alternatives. Does it build a *map*? Or does it build an amount-based, analog proportionate *meter* to represent sound location? [the latter two italics mine] […] To answer this, let's revisit the brain's mechanism for detecting *binaural* differences. [italics mine] […] [D]etecting a difference in sound arrival time across the two ears requires neurons that receive input from both ears. Slightly delaying the input from one ear relative to that of the other can make the two ears' signals, originating at the two eardrums at different times, arrive at a "comparison" neuron at about the same time. The comparison neuron responds better when the two inputs "line up"; that is when the sound's additional *physical* travel time to the "far" ear is precisely compensated by a matching excess of *neural* travel time from

> the "near" ear ... [...] So, does that make the code digital or analog? A map or a meter? Surprisingly, the answer seems to be that it depends.[14]

A simple answer to the concluding question could be: on apprehending a sound, or a letter, or a word, it might depend on a *digital code/map* being activated; when a melody, or a spoken phrase is apprehended, an *analogue*/a *meter* may be activated.

Back to basics

Can this be linked to where the difference between "binary opposition" and "binary difference" lies? In the search for similarities or differences, there remain a lot of ideas for future studies – on the physical side when it comes to frequency, wavelength, intensity etc.; on the neurobiological plane, not least in the use of the new brain-scanning methods, to map neuronal formations, transmitter substances, synaptic transmissions, network formations, etc. I will stop here and go to Ockham's razor – not to present the most economical description, but rather to present an idea: *For auditory sense it is about separating details before they are aggregated into configurations, for example, in the feature of a melody or a spoken phrase. For sense of sight it is about aggregating of configurations before separating into details*. Here I will insert a quote from Eric Kandel as support for my idea:

> From birth onward, infants are much more likely to look at faces than at other objects. [...] Modern [...] neurobiological studies have explained why human faces [...] are so special – they are specific Gestalt percepts. We perceive them as a unified whole as soon as our senses detect them. Instead of trying to process a face from a pattern of lines, [...] the brain uses a template-matching approach. It reconstructs the face from a more abstract, higher-order *figural primitive*: an oval containing two dots (for eyes), a vertical line between those dots (for the nose), and a horizontal line below them (for the mouth). (italics mine)[15]

14 Jennifer Groh, 2014, pp. 144–156.
15 Eric Kandel, 2012, pp. 287–288.

Do I have anything to add as a fundament for this proposal? I will go back to the origins of ontogenesis and to what is known about the child's development from foetus to newborn. As is made clear in my earlier books in Swedish, foetuses who during the pregnancy repeatedly have been listening to certain melodies and language chants show signs after birth of "auditory recognition" when they listen to the same melodies or language chants. The researchers who have reported these results are inclined to interpret the outcome as concerning the rhythmic pattern – "motor recognition" – rather than the melodies or language chants that the child is "auditorily recognizing". Why I put quotation marks around "recognizing", as readers will already know, is that I associate this performance with the ability of internal presentation that the innate child does not yet possess.

How should we understand what the innate child is showing? If we follow Steven Mithen's view, it can be explained by a transmission according to the "Hmmmmm model" – holistic, manipulative, multi-modal, musical, mimetic.[16] The elements I am prepared to embrace in Mithen's model are "holistic" in the sense of perceiving "rhythmic patterns"; "manipulative" because researchers have been preoccupied with asking for repeated presentations. However, I am hesitant about the "multimodal" being an asset to the innate child (see above); and about "musical" in the sense of being pitch loops. On the other hand, I embrace "mimetic" in the sense of being an ability to mimic. How can I justify what I accept and what I reject? In order to try to answer the question, I will start from the beginning, from an evolutionary and phylogenetic perspective.

The earliest and deepest inheritance of *Homo psychicus* is the rhythmic activity/biphasic cycle with expectation a priori as an evolutionary advantage. Throughout the mammalian chain we may observe manifestations of advanced and complex sensorimotor outliers – but also the emergence of specific organs like the elephant's trunk. When you see a chimpanzee carefully peeling a banana (or a squirrel a cone), you may already assume the two-hand grip as an evolutionary alternative. For all the way to two-handed *manipulation*, as shown in Chapter 4, bipedalism was the decisive

16 Steven Mithen, 2005.

step on that road; with that we as *Homo*, as collectors, scavengers and later as hunters with hunting gear. No chimpanzee "knows" anything about rhythm, nor can it stage-manage a rhythm. In that, Man is outstanding. Mithen is prepared to bet that the rhythmic dance was a first cult in a flock of *Homo sapiens* creatures – to pay tribute to the alien setting that gave luck in hunting? With some certainty, we can know something about these things in the light of archaeological finds and relics, eventually cave paintings (Chapter 5).

I would be willing to bet that there was a *sapiens* creature that, with a wooden bludgeon in his or her hand, discovered the rhythmic throbbing; with another *sapiens* creature who discovered what the throbbing of the first being could accomplish with his or her arms, hands and a wooden bludgeon in his hand, synchronously; and that the other being could follow and accompany with his or her legs and feet. Is there any neurobiological support for this proposal? I will quote Eric Kandel: "faces, hands, and bodies are so special [...]. We perceive them as a unified whole as soon as our senses detect them ..."[17] (which can also contribute to the understanding that the three-year-old's first drawing is a cephalopod). Additionally, applied to hearing, with its limited bandwidth, it had probably been sharpened into mastery when it came to making a distinction between sounds and calls from sources far beyond the domain of vision, but also with the support of the darkness of the night. But let's go back to my two *sapiens* creatures that I will expand with a third creature with the two others in the reach of the *eyes and ears*. When this creature with its wide bandwidth *sees*, with the aggregating – or to borrow a term from quantum mechanics "the entanglement" – visual system, before achieving a state of multimodality, the hand and arm patterns of the first creature follow, as in a wave movement, the foot and leg pattern of the second creature, the synchronous/now/in Augenblick contributes to the triggering first of surprise and immediately afterwards (at least embryonic) interest as amplifying both stimulus and response – creature one in relation to creature two – and the triad that was established in creature three. (Is this perhaps also the first moment/

17 Eric Kandel, 2012, p. 287.

embryonic state of multimodality?) But at the same time, also the rhythmic dance, all in one. Now there was also the prerequisite for a rapid transmission of the rhythmic dance to other creatures in the *sapiens* flock. With that, the first flock ritual also became an event in the world. Somewhere in Africa, or at about the same time in a few scattered places there. Moreover, afterwards affectively founded vocalizations were added to this. My more of a parenthetical thought about consonant calls is that they are a result of hearing's distinction of the slurping of the lips from the tip of the tongue meeting with the teeth. Is it the case that visual perception is basically symmetrical and that auditory perception is fundamentally asymmetrical?

If the events are as I describe them, they would already have been preceded by – in this I follow Dean Falk – a mother who puts the baby down on the ground in order to get hold of a hard-to-reach root in a bush (Chapter 5). The first vocalizing arc was struck between the crying baby and the comforting humming mother. Was it in this situation that "the third in the eye contact" was created (Chapter 3)? An eye contact between a mother and her baby where the mother fixates her interested gaze in her baby's gaze – and for the first time her baby responds with the affect of interest translated into his/her eye expression? Or vice versa, between a baby and his or her (allo) mother (see above)? If the answer to the questions is yes, semantics – and, if you wish, the whole meaning of life – would be grounded in earth. Is this the moment to which Ludwig Wittgenstein refers with his final words in *Tractatus Logico-Philosophicus*: "Whereof we cannot speak, thereof we must be silent"?[18]

But let's go back to hearing with its separating before aggregating and sight with its aggregating before separating. The auditory memory (hearing, an outgrowth on the evolutionary substantially earlier vestibularis) and the fate of the memory of vision does not apply to east and west: that never the twain shall meet (at the time when this was written, an extremely strained and accentuated military political relationship between the two prevailed). Through multimodality, hearing and sight in adults are in constant dialogue with each other. But without the hearing's separating feature – making a

18 Ludwig Wittgenstein, 1922/1974, p. 89.

difference? – followed by aggregating, there would be no music and no spoken language. And without sight's access to first aggregating and then separating, there would be no written language. I conclude with a beautiful quote by Falk:

> Music is not a useless spandrel [see Stephen Jay Gould, Chapter 4]. Rather, it contributed to making us the intelligent, vocal species we have become, and – especially when expressed as lullabies and playsongs – continues to benefit our children today. [...] Ultimately, if our ancestors' babies had not lost the ability to cling to their mothers, there would be no Mozart piano concertos and no Mona Lisa. No ballet, no Shakespeare, and no $E = mc^2$. (brackets mine)[19]

Perhaps hearing in its absolute origin was separating, with the purpose of assisting the vestibularis in the balancing act, not only to master gravity vertically and horizontally, but also with the aid of the sound from sources beyond the horizon; but even from the whole of space. But through sight and multimodality aggregating contributed, first, music and then eventually with spoken language as a brilliant extension (with written language as an additional extension). If so, the icon as a matter of both vision (the Gestalt) and hearing (the pitch loop) is hardly a mystery. If, with my sense modalities, I allow my final speculation to take some space, we would not have had syntax without the separation of hearing, or semantics without sight's aggregating of a mother's affect interest together with her baby's expression of interest, face-to-face, eye-to-eye, and vice versa. Could it even be the case that the earliest tendency of separation of hearing is the contribution of the senses to the *analytical* element of cognition – while the tendency for sight's aggregating (before separating) is a contribution to the *synthesizing* element of cognition? What we know about evolution is that it initially used the recipe: what is, is there to be built on.

19 Dean Falk, 2009. The quotation is a compilation of sentences on p. 135 and p. 177.

Bibliography

Anderson, P. (1972). More is different. *Science, 4047*, 393–397.

Anzieu, D. (1985/2016). *The Skin-Ego*. London: Karnac.

Arieti, S. (1974). *Interpretations of Schizophrenia*, 2nd ed. London: Crosby Lockwood Staples.

Aristotle. (1994–2009). *Nichomachean Ethics*, Book I, Point 7. The Internet Classics Archive.

Bates, E., Benigni, L., Bretherton, L., Camaioni, L. & Volterra, V. (1979). *The Emergence of Symbols: Cognition and Communication in Infancy*. New York: Academic Press.

Bateson, G. (1972). *Steps to an Ecology of Mind: Collected Essays in Anthropology, Psychiatry, Evolution, and Epistemology*. New York: Random House/Ballantine.

Bejarano, T. (2011). *Becoming Human: From Pointing Gestures to Syntax*. Amsterdam/Philadelphia: John Benjamins Publishing Company.

Bénard, H. (1900). Les tourbillions cellulaires dans une nappeliquide. *Science Pure Application, 11*, 679–686.

Bender, L. (1952). *Child Psychiatric Techniques*. Springfield: Thomas.

Blaffer Hrdy, S. (1999/2000). *Mother Nature: Maternal Instincts and How They Shape the Human Species*. New York: Ballatine Books.

Blaffer Hrdy, S. (2009/2011). *Mothers and Others: The Evolutionary Origins of Mutual Understanding*. Cambridge, Massachusetts/London: The Belknap Press of Harvard University Press.

Brown, G. (1914). On the nature of the fundamental activity of the nervous centres; together with an analysis of the conditioning of rhythmic activity in progression, and a theory of evolution of function in the nervous system. *Journal of Physiology, 48*, 18–46.

Carlberg, I. (2008/2011). *Pillret* (*The Pill*). Stockholm: Norstedts.

Changeux, J.-P. (2002/2009). *The Physiology of Truth: Neuroscience and Human Knowledge*. Cambridge, MA, and London: The Belknap Press.

Chomsky, N. (1957). *Syntactic Structures*. The Hague: Mouton.

Chomsky, N. (1988). *Language and Problems of Knowledge: The Managua Lectures*. Cambridge, MA: The MIT Press.

Clarke, A. (2008). *Supersizing the Mind: Embodiment, Action, and Cognitive Extension*. Oxford: Oxford University Press.

Close, F. (2000/2013). *Lucifer's Legacy: The Meaning of Asymmetry*. Mineola, NY: Dover Publications.

Coats, S. (2006/2009). Developmental research on childhood gender identity disorder. In P. Fonagy, R. Krause & M. Leuzinger-Bohleber (eds), *Identity, Gender, and Sexuality 150 Years After Freud: Controversies in Psychoanalysis*, 103–131. London: Karnac.

Corballis, M. (2002). *From Hand to Mouth: The Origins of Language*. Princeton, NJ: Princeton University Press.

Cozolino, L. (2006). *The Neuroscience of Human Relationships: Attachment and the Social Brain*. New York: Norton.

Damasio, A. (1999/2000). *The Feelings of What Happens: Body, Emotion and the Making of Consciousness*. London: Vintage.

Darwin, C. (1859/2015). *On the Origin of Species by Means of Natural Selection, or the Preservation of Favoured Races in the Struggle for Life*. Darwin Online.

Darwin, C. (1872/2009). *The Expression of the Emotions in Man and Animal*. London: Penguin Books.

da Silva, E., Ellbin, S., Marne, E., Nilsson, A. & Svensson, B. (1999). Intrapsychic changes among juvenile delinquents in 1984–1998 as studied by PORT, Perceptgenetic Object-Relation Test. *Early Child Development and Care, 149*, 71–85.

Deacon, T. (2012). *Incomplete Nature: How Mind Emerged from Matter*. New York: Norton.

Dencker, J. & Wålinder, J. (2008). Vi behöver en ny psykiatri utan okunnigas inblandning (We need a new psychiatry without ignorant interference). *Läkartidningen, 105*, 421–422.

Dimberg, U. (1982). Facial reactions to facial expressions. *Psychophysiology, 19*, 643–647.

Dimberg, U., Thunberg, M. & Elmehed, K. (2000). Unconscious facial reactions to emotional facial expressions. *Psychological Science, 11*, 86–89.

Dyson, G. (2012). *Turing's Cathedral: The Origins of the Digital Universe*. New York: Pantheon Books.

Einarsson, J. (2013). *Språket och språkarna (Language and Language Users)*. Lund: Studentlitteratur.

Enquist, M. & Eriksson, K. (2007). *Tvärsnitt*, 2. <http://www.vr.se/hudmeny/tvarsnittnr22007>.

Falk, D. (2009). *Finding our Tongues: Mothers, Infants & the Origin of Language*. New York: Basic Books.

Fogel, A., King, B. & Schancker, S. (eds). (2008). *Human Development in the Twenty-First Century: Visionary Ideas from Systems Scientists*. Cambridge: Cambridge University Press.

Fonagy, P., Krause, R. & Leuzinger-Bohleber, M. (eds) (2006/2009). *Identity, Gender, and Sexuality 150 Years After Freud: Controversies in Psychoanalysis*. London: Karnac.

Foucault, M. (1976/1979). *The History of Sexuality*: Vol. 1. *An Introduction*. London: Allen Lane. In Swedish 2002.

Foucault, M. (1961–1972/2006). *History of Madness*. London: Routledge & Kegan Paul. In Swedish 2010.

Frances, A. (2013/2014). *Saving Normal: An Insider's Revolt against Out-of-Control Psychiatric Diagnosis, DSM-5. Big Pharma, and the Medicalization of Ordinary Life*. New York: William Morrow HarperCollins Publishers.

Frank, H. (1962). *Kybernetischen Grundlagen der Pädagogik*. Baden-Baden: Agis-Agis Verlag.

Freeman, V. (2000). *How Brains Make Up Their Minds*. New York: Columbia University Press.

Freud, S. (1893–1895/1966). *Project for a Scientific Psychology*. Standard Edition, 1. London: Hogarth Press.

Freud, S. (1900/1953). *The Interpretation of Dreams*. Standard Edition, 4 and 5. London: Hogarth Press. In Swedish 1996.

Freud, S. (1915a/1957). *Repression*. Standard Edition, 14. London: Hogarth Press.

Freud, S. (1915b/1957). *The Unconscious*. Standard Edition, 14. London: Hogarth Press. In Swedish 2003.

Freud, S. (1926/1959). *Inhibitions, Symptoms and Anxiety*. Standard Edition, 20. London: Hogarth Press. In Swedish 2001.

Freud, S. (1940/2010). *An Outline of Psycho-Analysis*. Mansfield Centre: International Psycho-Analytical Library, 35, Martino Publishing. In Swedish 2003.

Frölander, H-E. (2005). Affekternas betydelse för kommunikativ och kognitiv utveckling – en sammanfattning av professor Alf Nilssons artiklar i Psykologtidningen 2000 (The importance of affects for communicative and cognitive development – a summary of professor Alf Nilsson's articles in Psykologtidningen 2000). *Nyhedsbrevet, 2*, 14–17.

Gazzaniga, M. (2008/2009). *Human: The Science Behind What Makes Us Unique*. New York: Perennial.

Gazzaniga, M. (2015). *Tales from Both Sides of the Brain: A Life in Neuroscience*. New York: HarperCollins Publishers.

Goldstein, K. (1939). *The Organism: A Holistic Approach to Biology Derived from Pathological Data in Man*. New York: American Books.

Gopnik, A. (2009/2010). *The Philosophical Baby: What Children's Mind Tell Us About Truth, Love, and the Meaning of Life*. New York: Picador.

Groh, J. (2014). *Making Space: How the Brain Knows Where Things Are*. Cambridge, MA, and London: The Belknap Press of Harvard University Press.
Haugeland, J. (1998). *Having Thought: Essays in the Metaphysics of Mind*. Cambridge, MA: Harvard University Press.
Henshilwood, C. & d'Errico, T. (eds). (2002/2011). *Homo Symbolicus: The Dawn of Language, Imagination and Spirituality*. Amsterdam/Philadelphia: John Benjamins Publishing Company, 75–96.
Hofstadter, D. (2007/2008). *I Am a Strange Loop*. New York: Basic Books.
Humphrey, N. (1992/1999). *A History of the Mind: The Evolution and the Birth of Consciousness*. New York: Copernicus Springer.
Humphrey, N. (1998). Cave arts, autism and the evolution of the human mind. *Cambridge Archaeological Journal, 8*, 165–191.
Humphrey, N. (2011). *Soul Dust: The Magic of Consciousness*. Princeton, NJ: Princeton University Press.
Humphrey, P. (1997). How properties emergence. *Philosophy of Science, 64*, 1–17.
Isacsson, G. (2008). Psykiatri är en medicinsk specialitet – punkt! (Psychiatry is a medical speciality – period!). *Läkartidningen, 105*, 1120.
Iverson, J. & Thelen, E. (1999). Hand, mouth and the brain: The dynamic emergence of speech and gesture. In R. Núnez & W. Freeman (eds), *Reclaiming Cognition: The Primacy of Action, Intention and Emotion*, 19–40. Thorverton: Imprint Academic.
Izard, C. (1991/2004). *The Psychology of Emotions*. New York: Plenum Press.
Kandel, E. (2012). *The Age of Insight: The Quest to Understand the Unconscious in Art, Mind, and Brain, from Vienna to the Present*. New York: Random House.
Kant, I. (1771/1998). *Critique of Pure Reason*. Cambridge:
Cambridge University Press. In Swedish 2004.
Kant, I. (1790/1987). *The Critique of Judgment*. Indianapolis, IN: Hackett Publishing Company.
Kauffman, S. (1993). *The Origins of Order: Self-Organization and Selection in Evolution*. New York: Oxford University Press.
Kendon, A. (2009). Language's matrix. DOI 10.1075/gest.9.3.05ken.
Kernberg, O. (1975). *Borderline Disorders and Pathological Narcissism*. New York: Jason Aronson.
Kernberg, O. (1976). *Object Relations Theory and Clinical Psychoanalysis*. New York: Jason Aronson.
Kingdon, J. (2003). *Lowly Origin: Where, When, and Why Our Ancestors First Stood Up*. Princeton, NJ: Princeton University Press.
Knight, C. (2009). Language, ochre and the rule of law. In R. Botha & C. Knight (eds), *The Cradle of Language: Studies in the Evolution of Language*, 281–303. Oxford: Oxford University Press.

Kohut, H. (1971). *The Analysis of the Self.* New York: International Universities Press.

Kohut, H. (1977). *The Restoration of the Self.* New York: International Universities Press.

Koppitz, E. (1968). *Psychological Evaluation of Children's Human Figure Drawings.* New York: Grune & Stratton.

Kragh, U. (1955). *The Actual-genetic Model of Perception-Personality.* Lund: Gleerups.

Kragh, U. (1968). *Manual till DMT: Defense Mechanism Test.* Stockholm: Skandinaviska Testförlaget.

Kragh, U. & Smith, G. (eds) (1970). *Percept-genetic Analysis.* Lund: Gleerups.

Krause, R. (2006/2009). Drive and affect in perverse actions. In P. Fonagy, R. Krause & M. Leuzinger-Boheler (eds), *Identity, Gender, and Sexuality 150 Years After Freud: Controversies in Psychoanalysis*, 161–175. London: Karnac.

Kuhl, P. (2000). Language, mind, and the brain: Experience alters perception. In M. Gazzaniga (ed.-in-chief), *The New Cognitive Neurosciences*, 99–115. Cambridge, MA: The MIT Press.

Lambert, M. (ed.). (2013). *Bergin and Garfield's Handbook of Psychotherapy and Behavior Change*, 6th edition. New York: Wiley.

Laughlin, R. & Pines, D. (2000). The theory of everything. *Proceedings of the National Academy of Sciences, 97*, 28–31.

Leaky, R. (1994). *The Origin of Humankind.* New York: Basic Books.

LeDoux, J. (2002/2003). *Synaptic Self: How Our Brains Become How We Are.* New York: Penguin Books.

Leuzinger-Bohleber, M. (2006/2009). Psychodynamic and biographical roots of a transvestite development: Clinical and extra-clinical findings from a psychoanalysis. In P. Fonagy, R. Krause & M. Leuzinger-Bohleber (eds), *Identity, Gender, and Sexuality 150 Years After Freud: Controversies in Psychoanalysis*, 43–73. London: Karnac.

Lévinas, E. (1961/1979). *Totality and Infinity: An Essay on Exteriority.* Pittsburgh: Duquesne.

Lévi-Strauss, C. (1962/1966). *The Savage Mind.* Chicago: Chicago University Press.

Lévi-Strauss, C. (1995). *Myth and Meaning.* London: Routledge & Kegan Paul.

Libet, B. (1992). Models of conscious timing and the experimental evidence. *Behavioral and Brain Sciences, 15*, 213–275.

Lorenz, E. (1963). Deterministic nonperiodic flow. *Journal of Atmospheric Sciences, 20*, 130–141.

Lowenfeld, V. (1947). *Creative and Mental Growth.* New York: Macmillan.

MacNeilage, P. (2008). *The Origin of Speech.* Oxford: Oxford University Press.

Mahler, M. (1965). On the significance of the normal separation-individuation phase: With reference to research in symbiotic child psychosis. In M. Schur (ed.), *Drives, Affects, Behavior*: Vol. 2, 161–169. New York: International Universities Press.

Matte Blanco, I. (1975/1998). *The Unconscious as Infinite Sets: An Essay in Bi-Logic.* London: Karnac.

Matte Blanco, I. (1988/2005). *Thinking, Feeling, and Being: Clinical Reflections on the Fundamental Antinomy of Human Beings and World.* London: Routledge.

Mehler, J., Jusczyk, P., Lambertz, G., Halsted, N., Bertoncini, J. & Amiel-Tison, C. (1988). A precursor to language acquisition in young infants. *Cognition, 29*, 143–178.

Mitchell, P. & Ziegler, F. (2007). *Fundamentals of Psychology: The Psychology of Childhood.* Oxford: Psychology Press.

Mithen, S. (2005). *The Singing Neanderthals: The Origins of Music, Language, Mind and Body*. London: Weidenfeld & Nicolson.

Nazzi, T., Bertoncini, J. & Mehler, J. (1998). Language discrimination by newborns: Towards an understanding of the role of rhythm. *Journal of Experimental Psychology: Human Perceptual Performance, 24*, 1–11.

Nilsson, A. (1977). Adaptive and defensive aspects of the individual: A system approach to adaptation in relationship to a psychoanalytic anxiety model. *International Review of Psycho-Analysis, 4*, 111–123.

Nilsson, A. (1983/1994). *The Mechanisms of Defence within a Developmental Frame of Reference*. Lund: Gleerups.

Nilsson, A. (1992). Farsotens andra stadium (The second stage of the scourge). *Sydsvenskan*. 16 October, p. 4.

Nilsson, A. (1995). Differentiation between patients with schizophrenia and borderline disorders in the Percept-genetic Object-Relation Test, PORT. *British Journal of Medical Psychology, 68*, 287–309.

Nilsson, A. (2005). *Om Homo psychicus uppkomst: En biopsykologisk fantasi* (*On the Origin of Homo Psychicus: A Biopsychological Phantasy*). Lund: A. L. N. Förlag.

Nilsson, A. (2009). *Det omedvetna i nya perspektiv: Ett psykiskt system mellan hudens och känslans beröring* (*The Unconscious in New Perspectives: A Psychic System between the Touch of Skin and Feeling*). Stockholm/Stehag: Brutus Östlings Bokförlag Symposion.

Nilsson, A. (2010). Gråten skapade språket (Cry created language). *Dagens Nyheter. Essä*. 16 March, p. 8.

Nilsson, A. (2011). *Det främmande, det kusliga och tankens brist: Det analoga, det digitala och Homo psychicus* (*The Unknown, the Uncanny and the Deficit of Thought: The Analogue, the Digital and Homo Psychicus*). Mölndal: Brutus Östlings Bokförlag Symposion.

Nilsson, A. (2013). *Homo psychicus som enfald kontra tvåfald mot mångfald: Det analoga och det digitala som färdriktning mot en människoteori* (*Homo Psychicus as Onefold and Twofold: The Analogue and the Digital as a Route towards a Human Being Theory*). Höör: Brutus Östlings Bokförlag Symposion.

Nilsson, A. (2014). *Affekter, relationer, operationer – grunden för Homo psychicus: Mot en människoteori i dynamiken mellan det analoga och det digitala (Affects, Relations Operations – the Basis for Homo Psychicus: Toward a Human Being Theory in the Dynamics between the Analogue and the Digital).* Höör: Brutus Östlings Bokförlag Symposion.

Nilsson, A. & Svensson, B. (1999). *PORT – Percept-genetic Object-Relation Test: A Projective Method for Clinical Use. Manual.* Lund: Department of Psychology, Lund University.

Peirce, C. S. (1955). Logic as semiotic: The theory of signs. In J. Buchler (ed.), *The Philosophical Writings of Peirce*. New York: Dover Books.

Piaget, J. (1947/2001). *The Psychology of Intelligence*. New York: Routledge Classics.

Piaget, J. (1951/1972). *Play, Dreams and Imitation in Childhood.* London: Routledge.

Piaget, J. (1977/2001). *Studies in Reflecting Abstraction*. Sussex: Psychology Press.

Pinker, S. (1994/1995). *The Language Instinct: The New Science of Language and Mind*. London: Penguin Books.

Pinker, S. (2002). *The Blank State: The Modern Denial of Human Nature*. New York: Viking.

Pinker, S. (2013). *Language, Cognition, and Human Nature*. Oxford: Oxford University Press.

Polanyi, M. (1958/1962). *Personal Knowledge: Towards a Post-Critical Philosophy*. Chicago: Chicago University Press.

Reissland, N., Francis, B., Mason, J. & Lincoln, K. (2011). Do facial expressions develop before birth? <http://www.plosone.org/artcle/info:doi/10.1371/journal.pone.0024081>.

Rizzolatti, G. & Sinigaglia, C. (2006). *Mirrors in the Brain: How Our Minds Share Actions and Emotions*. Oxford: Oxford University Press.

Rizzolatti, G. & Sinigaglia, C. (2010). The functional role of the parieto-frontal mirror circuit: Interpretations and misinterpretations. *Nature Reviews Neuroscience, 11*, 264–274.

Savage-Rumbaugh, S. & Fields, W. (2011). The evolution and the rise of human language: Carry the baby. In C. Henshilwood & F. d'Errico (eds), *Homo Symbolicus: The Dawn of Language, Imagination and Spirituality*, 13–47. Amsterdam/Philadelphia: John Benjamins Publishing Company.

Savage-Rumbaugh, S. & Lewin, R. (1994). *Kanzi: The Ape at the Brink of the Human Mind*. New York: Wiley.

Schilder, P. (1964). *Contributions to Developmental Neuropsychiatry*. New York: International Universities Press.

Schore, A. (1994). *Affect Regulation and the Origin of the Self.* Hillsdale: Lawrence Erlbaum Associate.

Schore, A. (2003). *Affect Regulation and Disorders of the Self.* New York: Norton.

Schur, M. (1955). The ego in anxiety. In R. Loewenstein (ed.), *Drives, Affects, Behavior,* 67–103. New York: International Universities Press.

Searle, J. (1995). *The Construction of Social Reality.* New York: Free Press.

Searle, J. (1998/1999). *Mind, Language and Society: Philosophy in the Real World.* New York: Basic Books.

Searle, J. (2015). *Seeing Things as They Are: A Theory of Perception.* New York: Oxford University Press.

Sirios, S. & Shultz, T. (2003). A connectionist perspective of Piagetian development. In P. Quilan (ed.), *Connectionist Models of Developmental Process in Real and Artificial Networks,* 13–41. New York: Psychology Press.

Sonnby-Borgström, M. (2002). Between Ourselves: Automatic Mimicry as Related to Emphatic Ability and Patterns of Attachment. Doctoral thesis. Department of Psychology, Lund University.

Sullivan, H. S. (1962). *Schizophrenia as a Human Process.* New York: Norton.

Tallon, A. (2007). Can Levina's ethical metaphysics contribute to psychoanalysis? The case for and against. *The Psychoanalytic Review, 94,* 657–680.

Thompson, E. (2007). *Mind in Life: Biology, Phenomenology, and the Sciences of Mind.* Cambridge, MA: The Belknap Press of Harvard University Press.

Titelman, D., Nilsson, A., Estari, J. & Wasserman, D. (2004). Depression, anxiety, and psychological defense in attempted suicide: A pilot study using PORT. *Archives of Suicide Research, 8,* 239–249.

Titelman, D., Nilsson, A., Svensson, B., Karlsson, H. & Bruchfeld, S. (2011). Suicide-nearness assessed with PORT, the Percept-genetic Object-Relation Test: A replication and a reliability study. *Bulletin of the Menninger Clinic, 4,* 295–314.

Tomasello, M. (2008). *Origins of Human Communication.* Cambridge, MA: The MIT Press.

Tomasello, M. (2009). *Why We Cooperate.* Boston: Boston Review Books.

Tomkins, S. (1962). *Affect, Imagery, Consciousness:* Vol. I. *The Positive Affects.* New York: Springer.

Tomkins, S. (1963). *Affect, Imagery, Consciousness:* Vol. II. *The Negative Affects.* New York: Springer.

Tomkins, S. (1991). *Affect, Imagery, Consciousness:* Vol. III. *The Negative Affects: Anger and Fear.* New York: Springer.

Tomkins, S. (1992). *Affect, Imagery, Consciousness:* Vol. IV. *Cognition. Duplication and Transformation of Information.* New York: Springer.

Turing, A. (1936). On computable numbers, with an application to the Entscheidungsproblem. Read 12 November.

Varela, F., Thompson, E. & Rosch, E. (1991). *The Embodied Mind: Cognitive Science and Human Experience.* Cambridge, MA: The MIT Press.

Vinicius, L. (2010). *Modular Evolution: How Natural Selection Produces Biological Complexity.* Cambridge: Cambridge University Press.

von Bertanlanffy, L. (1968). *General System Theory: Foundations, Development, Applications.* New York: Braziller.

Vygotsky, L. (1933/2004). Imagination and creativity in childhood. *Journal of Russian and East European Psychology, 42*, 7–97.

Werner, H. (1948/1957). *Comparative Psychology of Mental Development.* New York: International Universities Press.

Whitaker, R. (2010). *Anatomy of an Epidemic.* New York: The Park Library Group. In Swedish 2014.

Wilden, A. (1972/1977). *System and Structure: Essays in Communication and Exchange.* London: Tavistock Publications.

Wilson, E. O. (2014). *The Meaning of Human Existence.* New York: Norton.

Winnicott, D. (1953). Transitional objects and transitional phenomena. *International Journal of Psycho-Analysis, 34*, 89–97.

Wittgenstein, L. (1922/1974). *Tractatus Logico-philosophicus.* London: Routledge & Kegan Paul.

Wolff, P. (1987). *Behavioral States and the Expressions of Emotions in Early Infancy. New Proposals for Investigation.* Chicago and London: The University of Chicago Press.

Author index

Subject index

FSC
www.fsc.org
MIX
Papier | Fördert
gute Waldnutzung
FSC® C083411

Kolibri 360 GmbH
Ferdinand-Jühlke-Straße 7
99095 Erfurt, Deutschland
produktsicherheit@kolibri360.de